Borna Müller

Bovine Tuberculosis in Africa

Borna Müller

Bovine Tuberculosis in Africa

Molecular Epidemiology and Diagnosis

Südwestdeutscher Verlag für Hochschulschriften

Impressum/Imprint (nur für Deutschland/only for Germany)
Bibliografische Information der Deutschen Nationalbibliothek: Die Deutsche Nationalbibliothek verzeichnet diese Publikation in der Deutschen Nationalbibliografie; detaillierte bibliografische Daten sind im Internet über http://dnb.d-nb.de abrufbar.

Coverbild: www.ingimage.com

Verlag: Südwestdeutscher Verlag für Hochschulschriften GmbH & Co. KG
Heinrich-Böcking-Str. 6-8, 66121 Saarbrücken, Deutschland
Telefon +49 681 37 20 271-1, Telefax +49 681 37 20 271-0
Email: info@svh-verlag.de

Approved by: Basel, Swiss TPH, Diss., 2008

Herstellung in Deutschland (siehe letzte Seite)
ISBN: 978-3-8381-3273-0

Imprint (only for USA, GB)
Bibliographic information published by the Deutsche Nationalbibliothek: The Deutsche Nationalbibliothek lists this publication in the Deutsche Nationalbibliografie; detailed bibliographic data are available in the Internet at http://dnb.d-nb.de.

Cover image: www.ingimage.com

Publisher: Südwestdeutscher Verlag für Hochschulschriften GmbH & Co. KG
Heinrich-Böcking-Str. 6-8, 66121 Saarbrücken, Germany
Phone +49 681 37 20 271-1, Fax +49 681 37 20 271-0
Email: info@svh-verlag.de

Printed in the U.S.A.
Printed in the U.K. by (see last page)
ISBN: 978-3-8381-3273-0

for Vanessa, Ivana and Damian.

Table of contents

Acknowledgments

The present thesis was conducted within the framework of several scientific research partnerships between the Swiss Tropical and Public Health Institute (Swiss TPH, formerly Swiss Tropical Institute, STI) and various African and European institutions. I would like to express my sincere thanks to the numerous people who helped me in so many ways during my doctorate studies.
First and foremost, I would like to acknowledge the great support I received from my supervisor and mentor at the Swiss TPH, Prof. Jakob Zinsstag and former and current members of his research group, who contributed to this work. Special thanks go to Prof. Marcel Tanner, director of the Swiss TPH. It was a great privilege to work in the stimulating environment of his institute.
Many thanks also to our collaborators in Chad, especially Dr. Richard Ngandolo who was in charge of the sample collection in Sarh and the laboratory work conducted at the Laboratoire de Recherches Vétérinaires et Zootechniques (LRVZ) in Farcha, N'Djaména and Dr. Colette Diguimbaye-Djaïbe, deputy director of the LRVC and in charge of the tuberculosis laboratory.
I thank Prof. Bassirou Bonfoh, director of the Centre Suisse de Recherches Scientifiques (CSRS) in Abidjan, Ivory Coast for coordinating and planning many of our research activities conducted in Africa and especially for his contributions to our work in Mali. I would also like to thank all collaborators and friends of the Laboratoire Centrale Vétérinaire (LCV) in Bamako, Mali.
I'm indebted to Prof. Rudovick Kazwala from the Sokoine University of Agriculture (SUA) in Morogoro, Tanzania, for a very fruitful collaboration and Dr. Naima Sahraoui, who directed our work on the molecular characterization of mycobacterial strains from Algerian cattle. I very much appreciated her dedication to this project and her efforts to establish a successful collaboration between our research group and her partners at the University of Saad Dahlab, Blida and the University center of El-Tarf, in Algeria.
My sincere thanks also to Prof. Ayayi Justin Akakpo and his collaborators from the Ecole Inter-Etats des Sciences et Médecine Vétérinaires (EISMV) in Dakar, Senegal for their support during my work in Senegal.
I am particularly indepted to Prof. R. Glyn Hewinson from the Veterinary Laboratories Agency (VLA) in the UK for a remarkably fruitful partnership, which contributed substantially to my scientific education. I'm enormously thankful to Dr. Noel H. Smith from the VLA and the University of Sussex for having been a

great teacher and for making science so exciting. His intellectual contributions undoubtedly constitute the backbone of my thesis. I would also like to particularly thank Dr. Stefan Berg for having supervised much of the molecular typing work and for being a good fellow throughout my time in the UK.

All through my work I received often much needed support (particularly when times were tough) from my parents Pierrette and Meini, my sister Carla, my friends and especially my wife Vanessa. I sincerely thank them for all their help, love and understanding.

Financial support

This work has received financial support from the Swiss National Science Foundation (project no. 107559) and was partially supported by the Wellcome Trust Livestock for Live Initiative and Prionics AG.

List of tables

List of figures

Abbreviations

AIDS	Acquired immune deficiency syndrome
AFB	Acid fast bacilli
APC	Antigen presenting cell
BCG	Bacille Calmette-Guérin
bp	base pairs
BTB	Bovine tuberculosis
CI	Confidence interval
CMI response	Cell mediated immune response
CSSI	Centre Suisse de Support en Santé International
dNTP	Deoxyribonucleoside triphosphate
DNA	Deoxyribonucleic acid
DR	Direct repeat
EISMV	Ecole Inter-Etats des Sciences et Médecine Vétérinaires
ELISA	Enzyme-linked immunosorbent assay
ETR	Exact tandem repeats
FCFA	Franc de la Communauté Financière Africaine (currency)
FPA	Fluorescence polarization assay
HIV	Human immuno-deficiency virus
IFN-γ	Interferon gamma
IL	Interleukin
IS	Insertion sequence
LCV	Laboratoire Central Vétérinaire
LRVZ	Laboratoire de Recherches Vétérinaires et Zootechniques
LSP	Large sequence polymorphism
MIRU	Mycobacterial interspersed repetitive units
MTBC	*Mycobacterium tuberculosis* complex
MVA	Modified vaccine virus Ankara
NALC	N-acetyl-L-cysteine
NK cells	Natural Killer cells
NTM	Non-tuberculous mycobacteria
OIE	Office International des Epizooties
OR	Odds ratio
PCR	Polymerase chain reaction
PPD	Purified protein derivative

PPD-A	PPD of *M. avium* origin
PPD-B	PPD of *M. bovis* origin
RD	Region of difference
RNI	Reactive nitrogen intermediates
ROI	Reactive oxygen intermediates
SICCT	Single intra-dermal comparative cervical tuberculin
SNP	Single nucleotide polymorphism
STI	Swiss Tropical Institute
SUA	Sokoine University of Agriculture
TNF-α	Tumor necrosis factor alpha
TB	Tuberculosis
USD	US Dollar
VNTR	Variable number of tandem repeats
WHO	World Health Organization

Part I

General introduction and research approach

A review on bovine tuberculosis with a focus on sub-Saharan Africa

Introducing *Mycobacterium bovis*

Mycobacterium bovis is the causative agent of bovine tuberculosis (BTB) and belongs to the *Mycobacterium tuberculosis* complex (MTBC) of bacterial strains [1-3]. The MTBC is a group of very closely related pathogens, which can cause tuberculosis disease with similar pathology in a variety of mammalian species [1]. The most prominent member of the MTBC is *M. tuberculosis*, the principle causative agent of tuberculosis in humans, causing each year more than 1.5 million deaths and having experienced a recent re-emergence through the advent of HIV/AIDS and the appearance of multi drug resistant strains [4].

BTB is a chronic, generally respiratory disease, which is clinically difficult to diagnose although emaciation, loss of appetite, chronic cough and other signs of pneumonia could be symptoms developing at relatively late stages of the infection in cattle [5]. Especially in developing countries, clinical forms of many other chronic, emaciating diseases, like African trypanosomiasis, chronic contagious bovine pleuro-pneumonia (CBPP) or chronic multiparasitism, are difficult to be distinguished from BTB. BTB pathology is characterized by the formation of granulomatous lesions, which can within the course of the disease regress or exhibit extensive necrosis, calcify or liquefy and subsequently lead to cavity formation [6,7]. During meat inspection procedures on cattle carcasses in slaughterhouses, tuberculous lesions are primarily found in the upper and lower respiratory tract and associated lymph nodes [6]. However, the bacteria can also develop a systemic infection, disseminate within its host and affect other organs [8].

M. bovis is a slow growing, facultative intracellular, aerobic and gram-positive bacterium with a dysgonic colony shape when cultured on Löwenstein-Jensen medium [9]. As all *Mycobacterium* spp., *M. bovis* has an unusual cell wall surface structure characterized by the dominant presence of mycolic acids and a wide array of lipids [10]. This waxy lipid envelope confers an extreme hydrophobicity, which renders the bacteria acid- and alcohol-fast, a feature that can be exploited to identify mycobacteria via the Ziehl-Neelsen staining technique [11]. The mycobacterial surface lipids also have a potent biologic activity and are thought to play a crucial role in pathogenesis [10].

M. bovis can be identified on the basis of specific biochemical and metabolic properties. E.g., *M. bovis* requires pyruvate as a growth supplement, is negative for niacin accumulation and nitrate reduction, shows microaerophilic growth on Lebek medium and is generally resistant to pyrazinamide [9,12]. In contrast, *M. tuberculosis* does not require pyruvate as a growth supplement, is positive for niacin accumulation and nitrate reduction, shows aerophilic growth on Lebek medium, and is usually not mono-resistant to pyrazinamide [9,12]. However, the unequivocal validity of these characteristics is challenged by several studies [9,13].
Different molecular markers and techniques have been discovered and developed in the past that allow the unambiguous identification and differentiation of *Mycobacterium* spp. and the members of the MTBC [14-18].

The burden and epidemiology of *Mycobacterium bovis* infections

The burden of BTB

BTB is primarily of economic importance as it can have a considerable direct effect on milk and meat production and animal reproduction [19]. Moreover, national and international trade and other economic sectors may be indirectly affected by the disease [19]. BTB can also infect wildlife and thus have unpredictable consequences for entire ecosystems. E.g., in the southern region of the Kruger National Park in South Africa, 38% of the buffalos are infected with strains of *M. bovis*, originally introduced from domesticated cattle [20]. Carnivores such as lions, cheetah and leopards, feeding on infected animals are frequent spillover hosts [20]. Moreover, wildlife reservoirs of *M. bovis* hamper disease eradication schemes in several countries [21,22]. BTB also bears a zoonotic potential and it is of public health concern [23,24]. Infections of humans with *M. bovis* are rare in most countries that successfully apply disease control measures. However, zoonotic transmission of *M. bovis* may be frequent in countries where the disease is enzootic in cattle. In a recent study in the San Diego region of California, USA, *M. bovis* infections accounted for 45% of the tuberculosis cases in children [25]; most of these children were of Hispanic origin with ties to Mexico, where BTB is prevalent. Importantly, persons with *M. bovis* infections were 2.6 times more likely to die during treatment than persons with *M. tuberculosis* infections [25]. The poor, especially in developing countries, are thought to be at highest risk to contract zoonotic tuberculosis [26] and also the observed higher susceptibility of HIV-infected persons to *M. bovis* infections is of major concern [27].

The problems associated with BTB are of particular importance for many countries in Africa and especially the arid and semi-arid zones, where more than 50% of all African cattle, sheep and goats are raised [28] and where the livelihood of millions of people is based on livestock farming [26]. According to Thornton et al., there are an estimate of 556 million poor livestock keepers in the world with 30% of them living in sub-Saharan Africa and being most severely affected by the consequences of BTB [26,29]. However, data on disease prevalence in cattle or wildlife or the frequency of zoonotic transmission is generally scarce. This is mostly due to the absence of disease surveillance, insufficient laboratory capacity and the lack of veterinary expertise and may lead to a general underestimation of the disease prevalence in these regions [5].

BTB epidemiology

M. bovis can infect most mammalian species although bovids and especially cattle are the main hosts [30]. Transmission between animals is mostly thought to occur by inhalation of contaminated aerosol [30]. Evidence for a generally respiratory route of infection has first come from studies of the tuberculous lesion distribution in cattle. Most commonly, the upper and lower respiratory tract and associated lymph nodes are affected by lesions and the minimum dose required to establish disease in artificially infected cattle appeared to be 1000 times less for the respiratory route than for the oral route [6]. However, infection can also occur via the gastro-intestinal tract when animals ingest contaminated food, water or milk. In this case, lesion distribution is characterized by the presence of mesenteric lymph node lesions [30]. In a study in Ethiopia, mesenteric lymph node lesions were more often found in grazing animals compared to animals kept indoors [31]. Therefore, infection via the gastro-intestinal tract may be more important in cattle subjected to extensive livestock production systems as they are commonly observed in Africa. Because close contact between animals is relatively rare in extensive production systems, aerosol transmission of *M. bovis* is most likely occurring at water points, or when animals are gathering at night for protection from predators or during the daytime under trees when seeking shelter from the sun [5]. Different routes of transmission may take place if tuberculosis infection becomes systemic and other organs such as the urinary tract or the mammary glands become involved [8,30]. Especially the latter can be responsible for early infections in calves [30]. Moreover, consumption of contaminated milk represents the most important route of zoonotic tuberculosis transmission although disease communication can be easily prevented by milk pasteurization [5,32].

Commonly found risk factors for tuberculosis disease in cattle are close contact of animals such as in intensive livestock production systems [5,31-33], increasing herd size [33] and increasing age [34-36]. Another important risk factor is the contact or proximity of cattle and wildlife reservoirs of *M. bovis*, which obstruct disease eradication schemes in a number of countries. E.g., in the UK the Eurasian badger (*Meles meles*) represents a maintenance host for *M. bovis* [37]. White-tailed deer (*Odocoileus virginianus*) has been identified as a disease reservoir in Michigan, USA [38], the brush tail possum (*Trichosurus vulpecula*) is a disease reservoir in New Zealand [39,40] and the African buffalo (*Syncerus caffer*) has been identified as a main reservoir host for *M. bovis* in southern Africa [21]. Recent work also indicates an important influence of host genetics on disease susceptibility. In this respect, a report from Ethiopia described lower disease prevalence in African cattle breeds compared to exotic animals [31,41,42]; but also differences between distinct zebu breeds have been described in Chad [43].

The main risk factors for *M. bovis* infections in humans are poverty, malnutrition, HIV-infection, the consumption of raw milk and close contact to livestock [5,32].

Distribution and prevalence of bovine tuberculosis in sub-Saharan Africa

BTB has been largely eradicated in the industrialized world with the exception of a few countries in which the presence of a wildlife reservoir obstructs disease elimination even though extensive control measures are applied [30]. However, it is well known that BTB is present in many developing countries [32]. Nevertheless, due to the absence of disease surveillance and control, little accurate information is available on the prevalence and distribution of *M. bovis* infections.

Within a Wellcome Trust funded project, we have recently established a network of scientists and veterinary authorities from major livestock producing countries in Africa in order to discuss the problem of BTB and promote appropriate interventions. The first workshop with participants from West Africa was conducted in June 2007 in Bamako, Mali, where the representatives of each country also conveyed detailed information on the situation of BTB in their country. The compiled information obtained from this BTB network meeting and published data from peer reviewed journals on BTB in sub-Saharan Africa shall be presented below. Information presented at the BTB network meeting that could not be found in peer reviewed journals is referenced as a personal communication (pers. comm.); all presentations are downloadable from the African BTB network website (http://www.africa-btb.net/).

Senegal and Gambia

Previously performed studies on the prevalence of BTB in Senegal have shown a relatively low occurrence of the disease with most of the cases being identified in animals originating from Mali (M. Mbengue, pers. comm.; [44]). In a recent study of the International Trypanotolerance Centre, also a rather low occurrence of tuberculosis disease was detected in cattle from the Gambia [45].

Mali

In a recent study of Sidibé et al. in 36 herds and 1087 animals in the peri-urban region of Bamako, a reactor prevalence of 19% was reported, using the single intra-dermal comparative cervical tuberculin (SICCT) test. The herd prevalence was at 94% [46]. In our own study at the Bamako abattoir, the apparent lesion prevalence was only at 2%; however, most of these animals were likely originating from the pastoral areas and were raised in an extensive livestock production system [47].

Burkina Faso

In a slaughterhouse study in Bobo-Dioulasso in 1995, an average lesion prevalence rate of 4% was reported and *M. bovis* was isolated from 38 of 100 animals with gross visible lesions [48]. In addition, one isolate was identified as a strain of *M. tuberculosis.* [48]. An abattoir study in Ouagadougou in 2005, reported a prevalence of 2% (G. Poda, pers. comm.) and other studies in Burkina Faso published prevalence rates of up to 28% (G. Poda, pers. comm.). In a study conducted in the region of Bobo-Dioulasso, of six herds tested by SICCT, only one appeared to be free of reactors [49]. Of 199 animals slaughtered, 38 showed gross visible lesions of which 20 presented a positive culture for pathogenic mycobacteria [49].

Niger

An abattoir survey in Shaki revealed an apparent lesion prevalence of 13% [45]. In one study, among 174 isolated MTBC strains from humans, none was of *M. bovis* [50], but an increase in the prevalence of extra-pulmonary tuberculosis cases in humans has recently been observed (I. Maikano, pers. comm.).

Ghana

In 1983 and 1986, surveys in cattle using the single caudal fold tuberculin test revealed reactor prevalence rates between 0% and 2% in northern Ghana [45]. However, in more recent studies using SICCT testing, a reactor prevalence of 14% was detected in the area of Dangme and Great Accra, a prevalence of 14% was also detected in Dodowa district, a 19% reactor prevalence was detected in Ningo sub-district [45]. Prampram and Osudoku districts recorded both an 11% reactor prevalence [45]. Of 64 MTBC strains isolated from human tuberculosis patients at the Korle-Bu teaching hospital between January and July 2003, 3% of the strains were identified as *M. bovis* by biochemical methods [51].

Nigeria

In a survey conducted in Ibadan by Cadmus et al., strains of MTBC isolated from human tuberculosis patients and from slaughter cattle carcasses were subjected to molecular typing [52]. Of altogether 60 MTBC strains isolated from humans, three (5%) were identified as strains of *M. bovis*, six (10%) as strains of *M. africanum* and the rest as *M. tuberculosis*. Of 17 MTBC strains isolated from cattle, two were identified as *M. tuberculosis*, with one of them showing a spoligotype pattern that was also detected in strains of *M. tuberculosis* isolated from humans. Altogether, 15 of 17 strains from cattle were identified as *M. bovis* and the spoligotype patterns of all these strains characteristically lacked spacer 30 in the standard spoligotyping scheme [53]. One of the spoligotype patterns detected in strains from cattle was also found in a strain of *M. bovis* from humans, thus indicating the zoonotic transmission of strains of *M. bovis* from animals to humans in Nigeria [52]. Evidence for zoonotic transmission of *M. bovis* has also come from other studies [45,54,55]. The prevalence in cattle has been assessed by a number of studies in several regions using different methods; it ranged between 1% and 13% but appeared to be higher in the southern area of the country (S.I.B. Cadmus, pers. comm.). *Mycobacterium* spp. were also successfully isolated from milk (S.I.B. Cadmus, pers. comm.; [45]).

Chad

Schelling et al. reported a reactor prevalence of 17% for SICCT testing of animals from 13 herds in the Chari-Baguirmi and Kanem region of Chad [56]. Delafosse et al. investigated the prevalence of BTB in cattle herds of the Abéché region and reported a positive reaction to bovine tuberculin in 1% of the animals and 12% of the herds examined. Positive reaction to avian tuberculin was reported for 2% of the animals and 18% of the herds [57]. A slaughterhouse study at the N'Djaména abattoir revealed an apparent lesion prevalence of 7% [43]. Interestingly, strains of *M. bovis* were significantly more often isolated from the Mbororo zebu breed compared to the Arab zebu breed and it was hypothesized that this observation may have been due to a differential susceptibility of the two breeds to *M. bovis* infections [43]. With few exceptions, the spoligotype pattern of the strains isolated from Chadian cattle characteristically lacked spacer 30 as observed in Nigeria [43]. Prior to this study, it was believed that most of the gross visible lesions observed in cattle carcasses at N'Djaména abattoir were due to bovine farcy, which is caused by *M. farcinogenes* (C. Diguimbaye-Djaïbe, PhD thesis). Indeed, in addition to *M. bovis*, several different species of non-tuberculous mycobacteria (NTM) were isolated in a large proportion of animals with lesions [58]. In our study at Sarh abattoir in southern Chad, gross visible lesions were found in 11% of the sampled slaughter animals [59]. Using a Bayesian modeling approach, the true prevalence of BTB could be estimated at 8% [60]. Our results suggested that 72% of the suspected tuberculosis lesions detected during standard meat inspections were due to other pathogens than *M. bovis* [60].

Cameroon

Njanpop-Lafourcade et al. reported the molecular characterization of *M. bovis* strains isolated from cattle in three different provinces of Cameroon [61]. The spoligotype pattern of all the strains lacked spacer 30 as described for strains from Chad and Nigeria, indicating the presence of possibly one single clonal complex of strains of *M. bovis* in the Central African region. On the basis of the similarity of spoligotype patterns of strains from France and Cameroon and the longstanding history of cattle importation from France, the authors stated the possibility of an introduction of *M. bovis* from France into Cameroon [61].

Sudan

In a study conducted by Sulieman & Hamid in 2002, altogether 120 animals exhibiting caseous lesions were sampled from different abattoirs in Sudan. Microscopic examination detected acid fast bacilli (AFB) in specimens from 64 animals [62]. Bacterial growth was detected in cultures of lesions from 54 animals. Strains of *M. bovis* could be isolated from 25 animals. Moreover, 21 strains of *M. farcinogenes*, 4 strains of *M. tuberculosis* and 1 strain of *M. avium* was identified [62].

Ethiopia

In a slaughterhouse study conducted from November 2001 to April 2002 at Addis Ababa abattoir by Asseged et al., a lesion prevalence of 2% was reported for cattle if a detailed post-mortem examination was applied [63]. A retrospective analysis of meat inspection records from 1992 to 2001 also revealed an annual increase of whole carcass condemnations by 0.34% [63]. A similar study at Hossana municipal abattoir detected a lesion prevalence of 5% if a detailed post-mortem examination was applied [64]. Other studies in Ethiopia revealed a higher prevalence of BTB in cattle kept indoors compared to free-grazing animals [31] and a higher susceptibility to *M. bovis* infection of exotic Holstein *Bos taurus* cattle compared to local zebu cattle [42]. Moreover, a study by Ameni et al. revealed a better performance of SICCT in Ethiopia if the cut-off value for positive test interpretation was lowered from > 4 mm (OIE standard cut-off) to > 2 mm [65]. The spoligotype pattern of 17 strains of *M. bovis*, isolated from a herd with a high prevalence of BTB was identical for all animals and recently published [66]. Berg et al. provided a comprehensive investigation on BTB in Ethiopia and showed a widespread distribution of the disease at an average prevalence of approximately 5% [67].

Uganda

Oloya et al. reported an SICCT reactor prevalence of 2% and a herd prevalence of 51% for cattle from different pastoral herds from Karamoja region and Nakasongola district [36,68]. Also, molecular analysis of bacteria isolated from lesions of cattle slaughtered at Kalerwe abattoir was performed [69]. The spoligotype patterns of the different strains of *M. bovis* allowed the identification of at least two distinct clonal groups of which one was characterized by the absence of spacers 3-7 [69]; interestingly, lack of spacer 3-7 was also observed in strains from Ethiopia [66]. Moreover, in the same study, multiple strains of NTM were isolated from cattle [69]. Asiimwe et al. have recently reported a lesion prevalence of 0.5% for slaughter cattle at a city abattoir [70]. AFB were detected in cultures of 20% of the animals with lesions; these consisted of 11 strains of *M. bovis* and 6 strains of NTM [70].

Democratic Republic of Congo

Mposhy et al. reported the isolation of *Mycobacterium* spp., identified by Ziehl-Neelsen staining and microscopy, from 8% of altogether 1500 animal carcasses slaughtered at Goma abattoir [71]. Intra-dermal tuberculin skin testing of 1000 animals from the same region subsequently revealed a reactor prevalence of 8%. Based on these results, the authors speculated, that zoonotic transmission of *M. bovis* may had accounted for the high prevalence of tuberculosis cases observed in the pastoralist communities of North-Kivu [71].

Kenya

Limited information is available on BTB in Kenya although neighboring countries report endemicity of the disease in cattle. Kang'ethe et al. recently reported a tuberculin reactor prevalence of 10% in cattle from urban and peri-urban areas in Nairobi [72] and previously, *M. bovis* was isolated from Kenyan baboons feeding on abattoir offal [73].

Tanzania

A survey in the zone of lake Victoria, identified an average of 0.2% of SICCT reactors in a total of 8190 cattle from 42 herds [74]. In another study in Dar es Salaam region 1% of the animals reacted positively to bovine tuberculin and the herd prevalence was at 10% [75]. In the Lugoba area, 1% of tuberculin reactors were found in cattle and herd prevalence was at 21% [75]. In a study, in the southern Highlands, 13% of the cattle reacted positively to SICCT and the herd prevalence was at 51% [34]. More recently the SICCT reactor prevalence in cattle from extensive and intensive production systems in the eastern zone of Tanzania was assessed. In the extensive pastoral production systems reactor prevalence was at 1% and in the intensive production systems at 2% [76]. A cross sectional study in 10'549 cattle from 622 herds in northern Tanzania revealed a low SICCT reactor prevalence of 1% and a herd prevalence of 12% [33]. Risk factors identified for BTB were age, herd size, keeping cattle inside at night, contact to wildlife and annual flooding of villages [33]. In other studies in Tanzania, *M. bovis* could be isolated from milk [77] and zoonotic transmission of *M. bovis* was shown by means of molecular typing techniques [78]. Also, one study described the isolation of strains of *M. bovis* from wildlife [79].

Malawi

In a country-wide cross sectional study in 1986, a random sample of 3481 cattle was tested for BTB using SICCT [80]. The overall reactor prevalence was at 4% but varied by district, breed and sex (males were more often affected than females); however, most of the differences in prevalence by district were explained by differences in breed and sex [80].

Zambia

In a cross sectional study in Monze district, 7% of altogether 2226 cattle tested, reacted positively in the tuberculin skin test and 33% of the herds contained reactors [81]. Phiri reported in 2006 that out of 32'717 cattle slaughtered at three abattoirs in the western Province of Zambia, 183 animal carcasses were totally condemned with 83% of these condemnations having been attributed to BTB [82]. In a cross sectional study in the Kafue basin, SICCT was used to test cattle from 106 herds. True herd level prevalence was estimated at 50%, however, with considerable differences between study areas [83]. In another study, 944 cattle from 111 herds from the livestock/wildlife interface areas of Zambia were tested. In Lochinvar and the Blue Lagoon area, the SICCT reactor prevalence was at 5% and 10%, respectively [35]. In Kazungula, an area outside the livestock/wildlife interface, SICCT reactor prevalence was only at 1% [35]. Other studies also reported tuberculosis in Zambian wildlife and specifically the Kafue lechwe (*Kobus leche*) [45,84]. More recently, different genotypes of strains of *M. bovis* isolated from cattle from the Kafue basin have been described [85].

South Africa

BTB is believed to have been initially introduced into South Africa by European settlers; moreover, additional imports of infected live cattle from several different countries may have contributed to shape the relatively heterogeneous *M. bovis* population structure that is observed today [20,86]. In 1969, South Africa has launched a BTB control program and hereafter, prevalence rates have considerably dropped in domesticated cattle to 0.4% in 1995, but occasional outbreaks still occur [86]. Animal tuberculosis infections have first only been detected in domesticated cattle, however, sharing of rangeland or frequent contact between domesticated animals and wildlife may have later on lead to the spread of the disease into new animal populations [20]. As a result, *M. bovis* has caused a dramatic epidemic in several wildlife populations with the African buffalo (*Syncerus caffer*) representing the most important maintenance host of *M. bovis* (see chapter on the importance of bovine tuberculosis in African wildlife further below) [20]. The introduction of *M. bovis* from domesticated cattle into wildlife is likely to have occurred in other parts of Africa, as well, but possibly remained largely undetected due to poor surveillance [79].

Madagascar

In a sample survey using intra-dermal tuberculin skin testing, Quirin et al. found reactor prevalence rates between 0% and 30% [87]. Molecular typing of strains of *M. bovis* isolated from cattle in different regions of Madagascar revealed a homogenous population structure and no geographical localization of strain types within Madagascar [88]. Spoligotype patterns of *M. bovis* strains suggested the presence of a single clonal complex, which was characterized by the absence of spacers 3-5 and 8-10 [88]. This pattern has so far not been found in *M. bovis* strains outside Madagascar. Matching molecular types of strains isolated from humans and animals also indicated the zoonotic transmission of *M. bovis* in Madagascar [88].

Economics of bovine tuberculosis

BTB affects the national and international economy in different ways.
The most obvious losses from BTB in cattle are direct productivity losses (reduced benefit), which can be categorized into slaughter and "on-farm" losses [89]. Slaughter losses comprise the cost of cattle condemnation and retention, with the loss from condemnation being essentially the purchased value of a slaughter animal and the loss from retention being a fraction of the value of a carcass. On-farm losses comprise the losses from decreased milk and meat production, the increased reproduction efforts and replacement costs for infected cattle [89].
Effects of BTB on cattle productivity have been previously reviewed by Zinsstag et al. [19]. Early studies in Germany estimated a decrease in milk and meat productivity in totally infected livestock of 10% ± 2.5% and 4% ± 2%, respectively [19]. Similar milk productivity losses were also estimated in studies from Canada, Spain and the U.S. [19,89,90]. The study in Canada also estimated the reproduction losses to one fewer calf in infected cows; the replacement losses were estimated at 15% for infected animals [19]. Gilsdorf et al., assumed a 20% reduction in calf weight and a 5% replacement cost for their cost effectiveness analyses in the U.S. [89]. More figures and references are

given in the studies of Zinsstag et al. [19], Gilsdorf et al. [89] and Bernues et al [90].

Apart from direct productivity losses, BTB has profound economic consequences for national and international trade. On an international scale, BTB affects access to foreign markets due to import bans on animals and animal products from countries where the disease is enzootic. This situation has also major implications for other economic sectors, which are linked to livestock production. Moreover, BTB can create inefficiencies in the world market as e.g. economically inefficient but disease-free exporting countries will receive more revenues than economically efficient countries, which cannot export animal products due to enzootic BTB [89].

Presence of the disease in wildlife has considerable economic consequences. Not only is disease eradication more difficult and costly but BTB can theoretically affect entire ecosystems with unpredictable impact on many areas of private interest such as e.g. tourism (also see next section [19,91]).

Finally, BTB has a zoonotic potential and can cause disease in humans. Depending on the rate of zoonotic transmission this can have important effects on the public health sector.

At our BTB network meeting in 2007 in Bamako, Mali, all participants from West-Africa emphasized the economic importance of BTB in their countries albeit comprehensive economic assessments were largely missing. Cadmus estimated the losses due to BTB in Nigeria at 12-20 million USD annually. Poda noted that previous studies in Burkina Faso have calculated a decrease in weight of 12-29% in infected cattle compared to uninfected cattle. A study from 2001 revealed that meat confiscations in 8% of all cattle slaughtered at abattoirs in Burkina Faso have lead to a total loss of meat corresponding to 22 million FCFA (G. Poda, pers. comm.). In Mali, approximately 41 tons of meat have been lost in 2006 because of meat confiscations (pers. comm. of representatives of the Direction Nationale des Services Vétérinaires, DNSV). Sahraoui estimated the slaughter losses in Algeria from carcass condemnations at 940 Euros / total condemnation. In Chad, the livestock production sector accounts for 15% of the gross domestic product and 30% of the exported goods. In 2006, approximately 6 tons of meat have been lost due to meat confiscations (C. Diguimbaye-Djaïbe and B.N.R. Ngandolo, pers. comm.). At

Port Bouët abattoir in Ivory Coast, the slaughter losses in 2006 were estimated at 20.5 million FCFA (L. Achi, pers. comm.).

The economic impact of BTB in Africa is exacerbated through a number of factors. First, the fast growing population, especially in urban areas, causes an increase in demand especially for diary products and meat and promotes the intensification of livestock production in peri-urban areas [19,92]. Importantly, intensive livestock production systems show generally a higher prevalence of BTB than extensive production systems [5,31,32,93]. Second, developing countries lack the financial resources for disease control. This leads to a vicious cycle in which increased poverty affects the means for disease control and vise versa. Third, wildlife reservoirs in Africa are difficult to control; also, contact between transhumant cattle herds and wildlife may be particularly difficult to prevent in Africa. Forth, African countries have little access to the international trade and sanitary measures in industrialized countries may be used for protectionist purposes. Fifth, the public and political awareness are very low [19]. Most of the representatives from the various West-African countries at the BTB network meeting in Bamako in 2007 reported a generally low public awareness, a low to fair awareness of the cattle holders and in the research community in Africa and generally a lack of governmental commitment.

Few studies exist that assess the costs and the benefits of BTB control schemes rather than only the costs and almost none that specifically investigate the situation in Africa [19]. Cost-effectiveness studies of BTB intervention and control schemes must consider the costs and benefits in all the sectors affected (animal health, public health, etc.) [19]. In a cost-effectiveness study of a mass-vaccination campaign in livestock against Brucellosis in Mongolia, the intervention was profitable if the benefits of public health, agriculture and the private households were considered [94]. This may particularly apply to BTB in Africa, where the disease affects many aspects of human life. However, classical test and slaughter campaigns as they have been carried out in industrialized countries may not be feasible in Africa due to the lack of financial resources; therefore, new strategies may have to be assessed for BTB control in Africa. Options considered by the OIE are to create disease free zones by sequestration of infected animals and repopulation with disease free animals. Particular emphasis for such disease free zones should be given to intensive peri-urban production systems and control measures may be coupled to the systematic pasteurization of milk produced for commercial use. However, such strategies may be unusable to target BTB in extensive livestock production systems.

The importance of *Mycobacterium bovis* infections in African wildlife

M. bovis infections in wildlife can affect the ecosystem; moreover, the disease constitutes a threat to endangered species and can hamper BTB eradication and control schemes in domestic cattle. Tuberculosis in wildlife poses serious difficulties for BTB control in the UK and Ireland where the badger (*Meles meles*) represents an important disease reservoir [95,96]. In New Zealand, the brush-tailed possum (*Trichosurus vulpecula*) is a maintenance host of *M. bovis* [39]. White-tailed deer (*Odocoileus virginianus*) has been identified as a reservoir for *M. bovis* in Michigan, USA [38]. In central and eastern Europe, *M. bovis* has been isolated from wild boars (*Sus scrofa*) [22,97,98]. In fact, since the awareness concerning the importance of wildlife reservoirs of *M. bovis* rose, an increasing amount of reports has been published that describe the isolation of the agent from a big variety of mammalian hosts in different regions throughout the world [20,22].

Relatively little is known about the importance of *M. bovis* infections in wildlife on the African continent. However, available data from southern Africa suggests that the prevalence of wildlife tuberculosis has reached a dramatic dimension [20]. The African buffalo (*Syncerus caffer*) is considered the most important reservoir of *M. bovis* [20]. Also, the Kafue lechwe (*Kobus leche*) has been identified as a maintenance host of *M. bovis* and other species such as the greater kudu (*Tragelaphus strepsiceros*) may represent additional disease reservoirs. Therefore, today, BTB in Africa is considered a "multi-species host-pathogen system" [20].

In the Kruger National Park in South Africa, a gradient of infection from south to north has been detected with 38% of the buffalos being infected in the southern region, 16% in the central region and 2% in the northern region [20]. Spillover to carnivores such as lions (*Panthera leo*), leopards (*Panthera pardus*), cheetahs (*Acinonyx jubatus*) and Hyenas (*Crocuta crocuta*) but also other animal species like the chacma baboon (*Papio ursinus*), and warthog (*Pharcovhoerus aethiopicus*) has been observed repeatedly [20,21].

Control measures to prevent transmission of *M. bovis* between domesticated cattle and wildlife include test and slaughter schemes in buffalo populations along with breeding programs to produce disease-free buffalo calves and the separation of domesticated animals and wildlife through electrified perimeter fences. However, it is believed that vaccination programs for cattle and wildlife would be the most effective control measures [20]. Unfortunately, the only currently available vaccine (the Bacille Calmette-Guérin strain; BCG) for cattle shows a rather low protective efficacy and further research in vaccine development is urgently needed [20].

In Zambia, mycobacteria have been isolated from free ranging Kafue lechwe and bushbuck (*Tragelaphus scriptus*) [84] and BTB prevalence in cattle has been found to be higher in wildlife/livestock interface areas [83]. In Tanzania, *M. bovis* was isolated from migratory wildebeest (*Connochaetes taurinus*), topi (*Damaliscus lunatus*) and lesser kudu (*Tragelaphus imberbis*) [79]. Antibody enzyme immunoassays have detected *M. bovis* specific antibodies in 4% of a collection of serum samples (N = 184) from Serengeti lions (*Panthera leo*), in 6% of the samples (N = 17) from buffalo (*Syncerus caffer*) and in 2% of the serum samples (N = 41) from wildebeest [79]. *M. bovis* has also been isolated from buffalo and warthog (*Phacochoerus aethiopicus*) in the Ruwenzori National Park in Uganda [99,100]. Interestingly, in the same study, NTM have also been isolated from lesions of both animal species [99,100]. In other preliminary studies on tuberculosis in Tanzanian and Ethiopian wildlife, only NTM have been isolated from animals with gross visible lesions (T. Lembo and R. Tschopp, pers. comm.).

More research is needed to assess the magnitude and importance of *M. bovis* infections in wildlife populations. However, considering the particular livestock production system in Africa, currently available information suggests that tuberculosis in wildlife could play an important role with regards to the persistence of the disease in certain areas and cattle populations.

Zoonotic infections of *Mycobacterium bovis* in Africa

In Europe, cases of *M. bovis* infections in humans were of great importance until implementation of BTB control programs in many European countries through animal test and slaughter campaigns and regular milk pasteurization schemes lead to a drastic decrease of zoonotic tuberculosis cases and to a lower prevalence of BTB in general [23,27]. Although sporadic tuberculosis cases due to *M. bovis* are still occurring in industrialized countries they are generally considered to be of minor importance [23,27]. However, several studies were able to show an association between *M. bovis* infections and exposure to infected animals and animal products [27,101,102]; another important risk factor is the co-infection with HIV [23,27,103]. Considering the high prevalence of BTB in cattle and HIV positive individuals in Africa, the close contact between cattle holders and animals and the fact that the regular pasteurization of milk is not well implemented in most African countries, these findings are of major concern with regards to the public health significance of BTB in Africa [5,32].

The contribution of *M. bovis* infections to tuberculosis cases in humans in Africa is not well known. This is primarily due to the lack of diagnostic laboratories with the means to distinguish strains of *M. tuberculosis* and *M. bovis* [5].

However, numerous population based studies on human tuberculosis in Africa did not find any evidence for transmission of *M. bovis* from animals to humans [104-110]. In one of our previous studies in Chad, 40 *M. tuberculosis* but no *M. bovis* strains could be isolated from human sputum samples [111]. Only *M. tuberculosis* strains could be detected in sputum samples from 46 tuberculosis patients at Dantec Hospital in Dakar, Senegal (own unpublished results). Table 1 lists a selection of population based studies in which *M. bovis* infection was detected by molecular or biochemical methods in African patients with pulmonary or extra-pulmonary tuberculosis. In three studies from Tanzania [33,112,113], in one study from Nigeria [55] and in one study from Uganda [114], *M. bovis* accounted for 20% or more of the MTBC strains isolated. In all other reports [51,52,54,115-122], The contribution of *M. bovis* to human tuberculosis cases was relatively low and comparable to prevalence rates found in European settings [32,123].

Interestingly, high proportions of *M. bovis* infections in humans were particularly detected in rural settings. This is most evident from the available data for Tanzania. In a recent study in Dar es Salaam, 147 *M. tuberculosis* but no *M. bovis* infections could be detected in human tuberculosis cases [105]; however, in the rural areas of Tanzania, *M. bovis* accounted for 20% of the MTBC infections detected (table 1). Similarly, strains of *M. bovis* were isolated in 30% of the MTBC infections found in pastoralist communities in rural Uganda but in less than 0.5% of the tuberculosis cases in urban Kampala (table 1). Relatively few information is available on tuberculosis in pastoralist communities in other African countries although they seem to be prone to *M. bovis* infections. Importantly, nomadic and transhumant pastoralists are often excluded from the public health services as they are difficult to reach. Conversely, nomadic and transhumant pastoralists are less likely to seek treatment as their and their families' livelihood largely depends on livestock farming. This difficultly allows the members of a pastoralist community to seek treatment in urban health centers for extended periods of time. Therefore, adapted, mobile health services would be required to provide effective tuberculosis treatment to mobile populations [124].

More research is needed to assess the importance of human tuberculosis due to *M. bovis* in Africa. However, currently available information suggests that zoonotic transmission may be of minor importance in urban areas but more frequent in rural areas.

Country	Study setting	Total MTBC	M. bovis	% M. bovis	Reference
Cameroon	15 district hospitals in Ouest province	455	1	0.2%	Niobe-Eyangoh et al., 2003
Djibouti	Unknown	85	1	1.2%	Koeck et al., 2002
Egypt	Six fever hospitals in different cities	67	1	1.5%	Cooksey et al., 2002
Ghana	Korle-Bu Teaching Hospital	64	2	3.1%	Addo et al., 2007
Guinea-Bissau	Unknown	229	4	1.7%	Källenius et al., 1999
Madagascar	Antananarivo, Antsirabe, Fianarantsoa, Mahajanga	400	5	1.3%	Rosolofo-Razanamparany et al., 1999
Nigeria	2 hospitals in Ibadan	60	3	5.0%	Cadmus et al., 2006
Nigeria	Lagos	91	4	4.4%	Idigbe et al., 1986
Nigeria	3 hospitals in Jos	50	10	20.0%	Mawak et al., 2006
Tanzania	4 districts of Manyara (Arusha) Region	34	7	20.6%	Cleaveland et al., 2007
Tanzania	Pastoralist communities in the North and South	38	7	18.4%	Kazwala et al., 2001
Tanzania	Rural and semi-rural districts of Arusha	34	7	20.6%	Mfinanga et al., 2004
Uganda	Kampala	344	1	0.3%	Asiimwe et al., 2008
Uganda	Kampala	234	1	0.4%	Niemann et al., 2002
Uganda	Pastoralists of transhumant areas in Karamoya	10	3	30.0%	Oloya et al., 2007

Table 1. A selection of reports on zoonotic tuberculosis infections in Africa.

Methods for molecular epidemiological investigations

The different genotyping techniques used for the molecular characterization of MTBC strains have been reviewed extensively [17,125-128].

Molecular typing of bacterial strains can offer general insights into the population structure of pathogens. In the case of MTBC, a high degree of genetic homogeneity of strains isolated from patients from a given setting indicates recent transmission of tuberculosis. Conversely, a heterogenic population structure, especially in a low prevalence area, could suggest that tuberculosis reactivation may be more frequent [126]. Moreover, molecular epidemiological tools can help to identify transmission chains, risk factors for tuberculosis infections [125] and they have been useful for the detection of laboratory cross-contaminations [128]. The most important techniques that have been used throughout our studies shall be presented below.

Spacer oligotyping (Spoligotyping)

Spoligotyping makes use of the variability of the MTBC chromosomal direct repeat (DR) locus for strain differentiation [53]. The DR region is composed of multiple well conserved direct repeats of 36 bp which are separated by distinct, non-repetitive spacer sequences of similar size [129]. In the standard spoligotyping scheme, a PCR with primers complementary to the DR-sequence is used to amplify all spacer sequences of a given strain. One of the two primers is labeled with a biotin marker. The PCR products are denatured and hybridized to a standard set of 43 oligonucleotides covalently linked to a membrane. These oligonucleotides correspond to 37 spacers from *M. tuberculosis* H37Rv and 6 additional spacers from *M. bovis* BCG P3. If any of these spacers are also present in an investigated strain, they will be amplified during the PCR and hybridized to the spacers on the membrane. The successful hybridization can be visualized by incubation with Streptavidin peroxidase (which binds to the biotin molecule), subsequent addition of a chemiluminescent Streptavidin peroxidase substrate and exposure to a light sensitive film [53]. Figure 1 provides a schematic overview of the spoligotyping procedure and gives an example of spoligotype patterns for different strains.

Compared to other molecular typing techniques, spoligotyping is not very discriminatory although its resolving power could be increased by inclusion of additional spacers in an extended spoligotyping protocol [130]. The technique has been proven to be useful for a first assessment of the MTBC strain diversity in many population surveys [102,131-134]. Spoligotyping also allows the differentiation of MTBC members; e.g., *M. bovis* strains characteristically lack spacers 3, 9, 16 and 39-43 [1]. Moreover, strain groups and families can be readily identified [135] and extensive databases (www.Mbovis.org and SpolDB4 [136]) have facilitated the comparison of strains isolated in different countries. It is assumed that the DR locus has a unidirectional evolution (spacers can only be lost and not reacquired) and spoligotyping can therefore be particularly useful for phylogenetic analyses [137].

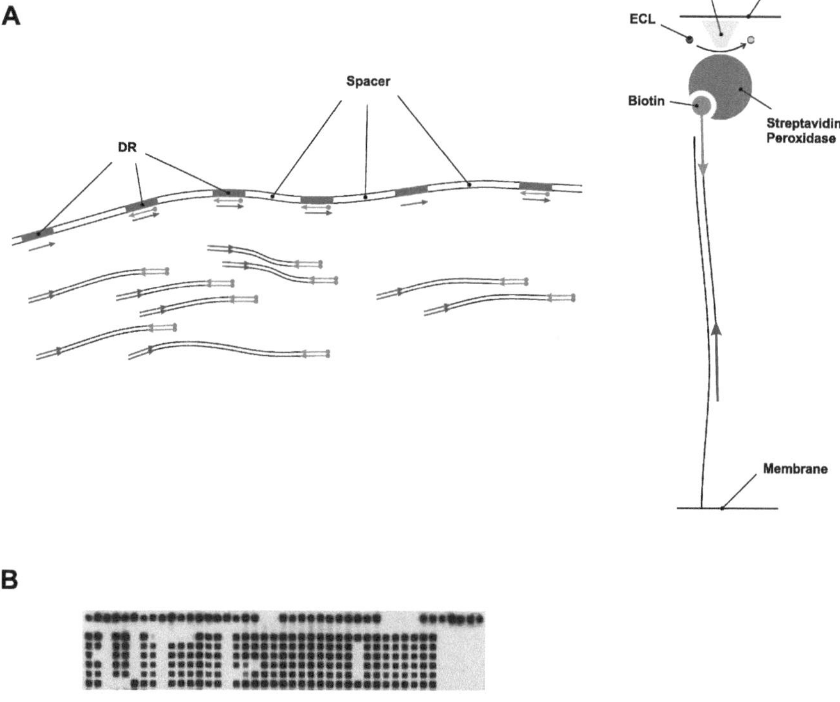

Figure 1: The spoligotyping method. A: Schematic overview of the spacer amplification during PCR on the left. Primers hybridize within the direct repeats (DR) with one primer being linked to a biotin molecule. PCR products of variable

length are generated. Detection of amplified spacers on the right. PCR products hybridize to complementary spacer oligonucleotides linked to a membrane. Streptavidin peroxidase associates with the biotin molecule and light is emitted upon addition of a peroxidase substrate [enhanced chemiluminescence (ECL) substrate]. B: Example of spoligotype patterns obtained for seven different samples and a negative control when 43 spacers are being represented on the membrane as described by Kamerbeek et al. [53].

Variable number of tandem repeat (VNTR) typing

Several genomic loci of MTBC with tandemly repeated minisatellite sequences have been identified [125]. In VNTR typing, the number of tandem repeats for a given locus is assessed by PCR amplification of the repeats with primers targeting their flanking regions. From the size of the PCR product, the number of repeats can be deduced. Tandem repeats showing variability between distinct strains have been identified in various studies [138-140] and by combining VNTR typing for different loci, a significantly higher discriminatory power can be achieved than for spoligotyping [141-143]. Moreover, VNTR typing is a relatively simple technique and the results consist of a digit code, which can be easily communicated and compared.

However, the technique is not well standardized [125] and large databases as for spoligotyping are missing. Therefore, VNTR typing is often used as an additional tool to sub differentiate clusters of strains with identical spoligotype patterns [1]

Large sequence polymorphism (LSP) analysis

LSPs generally refer to large genomic deletions or insertions. Large genomic deletions (also called regions of difference, RD) are widely used for the phylogenetic analyses of MTBC but are not appropriate for molecular epidemiological studies, due to a low mutation rate [1,144,145]. A major drawback of spoligotyping and VNTR typing is the possible occurrence of homoplasies; that is, the development of identical molecular types in unrelated strains [1]. Such events are very unlikely to occur for large genomic deletions unless an LSP is associated with repetitive and transposable elements. Moreover, because of the suspected absence of horizontal gene transfer in

MTBC, large genomic deletions cannot be reacquired and the direction of evolution can be determined without the use of an outgroup [1].
Large genomic deletions can be discovered through comparative genomic hybridization using bacterial genomic DNA and oligonucleotide microarrays. After specific LSPs or RDs have been identified, simple PCR protocols to rapidly assess their presence or absence in population surveys can be developed.

The evolution and population structure of *Mycobacterium bovis*

Evolution of M. bovis

The MTBC is defined as a complex of seven distinct bacterial species named *M. tuberculosis, M. canettii, M. africanum, M. pinnipedii, M. microti, M. caprae* and *M. bovis*. Although strains of the MTBC show a remarkable sequence similarity of at least 99.9% [1], the different bacterial species show a certain host tropism. E.g., *M. bovis* most commonly affects cattle, *M. tuberculosis* affects humans, *M. microti* is most frequently isolated from voles, etc. [1,137,145]. Nevertheless, spill over to other hosts has been observed for most of the bacteria [1,146,147]. Figure 2 shows the different clades of the MTBC and the phylogentically informative large genomic deletions (RDs) that allow their differentiation (adapted from Brosch et al. [145]). On the basis of these and other observations, it has been suggested, that the MTBC should rather be considered as a group of different "ecotypes", adapted to distinct hosts [137]. Generally, ecotypes can be described as the entirety of all individuals that occupy a specific ecological niche and that can be identified as a clustered group using specific genomic markers [148]. The identification of ecotypes of the MTBC and their defining molecular traits is still ongoing. In fact, host adaptation of MTBC strains may not necessarily be restricted to the species level but could be more specific [137,149]. Indeed, in a recent study, Gagneux et al. suggest a variable host-pathogen compatibility between distinct strains of *M. tuberculosis* and human hosts of different ethnicity [150].
Except for *M. canettii* [151], there is little evidence for horizontal gene transfer between other ecotypes of the MTBC [1]. Therefore, the population structure of MTBC can be considered as highly clonal if *M canettii* is excluded. The clonality of the MTBC and its ecotypes can be most evidently shown by the matching

phylogenetic trees that have been constructed using different molecular markers [1,144]. Also, the observation of a linkage disequilibrium between 12 independent microsatellite loci in strains of *M. tuberculosis* has suggested a clonal population structure [152]. The reason for the strict clonality of the MTBC could be due to an infrequent encounter of the different ecotypes [137], mutations in the recA gene or other barriers to recombination [1].

Because of the smaller genome size of *M. bovis* compared to *M. tuberculosis* and because no significant chromosomal regions are present in *M. bovis* but absent in all strains of *M. tuberculosis*, it can be inferred that *M. bovis* emerged from an *M. tuberculosis*-like ancestor. Therefore, it has been suggested that tuberculosis disease was first present in humans prior to animals [145]. This hypothesis has been a matter of controversial discussions [1]; however, recent work has added new evidence to this theory. In two independent studies, analysis of the VNTR and SNP profiles, respectively, of an extensive collection of MTBC strains allowed to differentiate two major clades of the MTBC. The first clade exclusively comprised strains of the human pathogen *M. tuberculosis*. However, the second clade comprised strains of *M. tuberculosis* and in addition all strains of *M. africanum, M. pinnipedii, M. microti, M. caprae* and *M. bovis* [2,3]. Parsimony thus suggests that the last common ancestor of both clonal groups already constituted a human host adapted pathogen [1-3]. Using Bayesian statistics, Wirth et al. estimated the age of the MTBC at 40'000 years. The calculated time of the MTBC emergence would thus coincide with the expansion of "modern" human populations out of Africa [2].

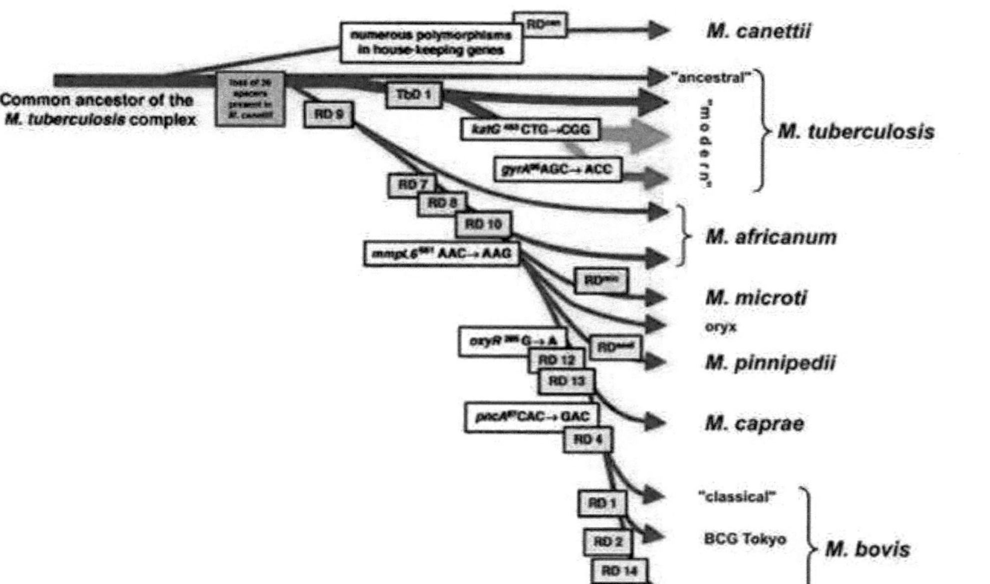

Figure 2: Evolutionary scenario of the MTBC (adapted from Brosch et al. [145]). Phylogentically informative large genomic deletions and single nucleotide polymorphisms (SNP) are indicated.

The population structure of M. bovis in the UK

Our knowledge about the global population structure of *M. bovis* is very limited and only few representative national surveys have been conducted, up to date. Most insights into the population structure of *M. bovis* came from molecular epidemiological studies from the UK, where BTB is still prevalent despite longstanding attempts to eradicate the disease [153]. The BTB control program in the UK started in 1935 and became highly successful in the 1960s and 70s when very low tuberculin reactor incidences were reached. However, incidence rates have again drastically increased since the late 1980s [153]. It has been shown that the Eurasian badger (*Meles meles*) represents a maintenance host for *M. bovis* in the UK and probably the establishment of this wildlife reservoir has accounted for the increase of the BTB prevalence in the UK [37].

The spoligotype patterns of most strains of *M. bovis* isolated from cattle in the UK characteristically lack spacers 6 and 8-12. Therefore, it has been suggested that the majority of strains found in the UK belong to a single clonal complex [1]. *M. bovis* population surveys in France, Spain and other countries of continental Europe indicated that the clonal complex existing in the UK is not present at high frequency outside the British Isles [1]. Using a combined approach of spoligotyping and VNTR typing, different molecular types of *M. bovis* could be discriminated at a high resolution [154]. It has been shown that these different types were geographically localized to specific regions where they have reached fixation or represent the majority of strains isolated [1]. Moreover, VNTR typing of strains showing the two most common spoligotype patterns, indicated that the frequency of the VNTR patterns could not be explained solely by random mutation and drift and may be best explained by a series of clonal expansions [154]. Due to the BTB control schemes in the UK, *M. bovis* experienced a severe bottleneck in the 1960s/70s. Therefore, the observed clonal expansion and the geographical localization of distinct molecular types could have been a result of population sampling as different clones invaded new ecological niches (founder effect) [154]. However, the clonal expansions could also be explained by selection [154]. It has been suggested that the strains of *M. bovis* in the UK may have been selected for their ability to maintain themselves in the badger population [154]. Alternatively, they may be more efficient in evading the BTB control scheme as the most common molecular types of *M. bovis* in the UK are different from the type of strain used for tuberculin production [154,155].

Immunology of bovine tuberculosis

Based on the predominance of tuberculous lesions in the respiratory tract and associated lymph nodes of diseased cattle, it was early believed, that BTB is transmitted from animals to animals through inhalation of infectious aerosols [6]. Further evidence for this route of transmission has come from a number of other studies [6]. Bacteria entering the respiratory tract, passing the mucociliary layer and gaining access to the alveolar space are thought to be phagocytized by macrophages, which may constitute the main cellular host for mycobacteria *in vivo* [7]. Following uptake into the phagosome, macrophages attempt to kill the bacteria by phago-lysosome fusion and acidification. However, mycobacteria are able to prevent lysosomal delivery by manipulating the host cell signal transduction pathways using an array of bacterial effector lipids and proteins [156]. Mycobacteria are therefore believed to reside and multiply primarily within the phagosomes before they eventually destroy the phagocytes. However, this assumption has been challenged by a recent study of van der Wel et al., describing the translocation of *M. tuberculosis* from phago-lysosomes to the cytosol of myeloid cells, from the second day after phagocytosis [157]. Although infected macrophages are not believed to act as the main antigen presenting cells (APC) they can trigger an immune response through the secretion of pro-inflammatory cytokines (e.g. TNF-α) and chemokines, leading to the recruitment of other phagocytes and lymphocytes to the lung [7,158].

The most important APCs in tuberculosis infection are probably dendritic cells, which also play a major role in modulating the host immune response. Mycobacteria may get access to the lung tissue through M-like cells in the bronchi. From there, they can penetrate to the underlying lymphoid tissues, get phagocytized by dendritic cells and transported to the draining lymph nodes, where immune responses are initiated [7]. The interaction of *M. bovis* with bovine dendritic cells leads to cell maturation and increases expression of surface molecules involved in T-cell interactions [159]. Moreover, altered cytokine profiles and especially the secretion of IL-12 and TNF-α are observed upon infection of bovine dendritic cells with *M. bovis* [159]. The secretion of these cytokines and mycobacterial antigen presentation on the surface of dendritic cells help triggering an adaptive cell mediated immune (CMI) response [158]. This CMI response is commonly known as the Th1-type immune response and characterized by the secretion of high levels of IFN-γ and IL-2 by Th1 CD4 T-cells [158]. The production of IFN-γ and IL-2 by Th1 cells can activate macrophages in order to become highly microbiocidial. Following activation, macrophages may be able to kill most of the mycobacteria within the phagosome by releasing increased amounts of hydrolytic enzymes, reactive oxygen intermediates and reactive nitrogen intermediates (including nitric oxide) [7,158]. Interestingly, activation of bovine dendritic cells enhances their ability to kill ingested *M. bovis* to a lesser extend and significant numbers of live bacilli are able to persist. Dendritic cells may therefore constitute a reservoir for pathogenic mycobacteria [7]. However, in an non-activated state, dendritic cells are more potent in killing ingested bacteria than macrophages [7].

Natural Killer (NK) cells and γδ T-cells are crucial in the early innate immunity against mycobacterial infections. They release IFN-γ when activated through the interaction with dendritic cells and thus contribute to the Th1 biased immune response [7]. Conversely, NK and γδ T-cells also play a role in fully activating dendritic cells [7]. Control of infection is ultimately dependent on the induction of a strong CMI response and granuloma formation, in which CD4 but also CD8 T-cells play a pivotal role. If the immune response against mycobacterial infection is strong enough to contain bacterial growth, active disease does not develop although infection may never be fully cleared. If the balance between the host's defenses and the persisting mycobacteria is tipped in favor of the pathogen, active disease occurs and the granuloma formation progresses [160]. At a late stage, BTB granulomas are characterized by extensive necrosis which can lead to liquefaction and cavity formation. Rapture of these cavities into the bronchi consequently allow aerosol spread of the bacteria [158]. In a study of Cassidy et al., microscopic lesions were observed in experimentally infected cows as early as seven days after inoculation of *M. bovis* [161]. Gross lesions were detected in the upper respiratory tract, in the lungs and the lymph nodes draining these areas at 14 days post-infection [161]. At the slaughterhouse, granulomatous lesions are most often detected in lymph nodes (especially mediastinal or bronchial). This is due to the fact that fluids in an animal together with activated macrophages eventually pass through the lymph nodes where the pathogens are filtered [158].

Although necessary for protection, the CMI response in tuberculosis infection can also contribute to the immuno-pathogenesis of tuberculosis [7]. The Th2-type immune response functions as a regulatory element to counteract and downregulate the pro-inflammatory CMI response, elicited by IFN-γ producing cells [7]. It is also associated with an increased humoral immune reaction and is mainly triggered by CD4 Th2 cells and the secretion of IL-4. It is believed that Th1-type immune responses wane towards late disease stages. In contrast, humoral Th2-type immune responses increase as disease and pathology progresses and bacterial load increases [162]. Although the factors that trigger the transition from a Th1- to a Th2-type immune response are not well known, it is thought that the CMI response is most relevant to determine protection; in contrast, the humoral response is considered detrimental [158].

Diagnosis of bovine tuberculosis and vaccine development

Diagnosis of BTB

Today, the two most widely used tests for the diagnosis of BTB are the single intra-dermal comparative cervical tuberculin (SICCT) test and the Bovigam® test (Prionics) [162]. Both are based on the detection of the early CMI response in tuberculosis infection. The SICCT test measures the delayed-type hypersensitivity response to intra-dermally injected bovine and avian tuberculin. Tuberculin or the so-called purified protein derivative (PPD) is a crude extract of water soluble proteins from a heat-treated culture of *M. bovis* (PPD-B) or *M. avium* (PPD-A) [162]. PPD-B and PPD-A are injected side by side into the skin of an animal's neck and skin swelling is measured 72 hours later. According to the OIE standard procedure, the test result is considered positive, if the difference in the increase of skin thickness at the site of PPD-B injection is more than 4 mm greater (> 4 mm) than the increase in skin thickness at the site of PPD-A injection [163].

The Bovigam® test measures the *in vitro* IFN-γ production of whole blood cells after exposure to PPD-B and PPD-A for 16-24 hours [162,164]. IFN-γ production is quantified by a sandwich enzyme-linked immunosorbent assay (ELISA). A large disadvantage of the Bovigam® test is the requirement to quickly process the blood samples soon after collection; however, the method appeared to be somewhat more sensitive than SICCT [162,164].

CMI response based diagnostic tests allow a relatively early detection of *M. bovis* infection in cattle. At late disease stages, the CMI response can wane as opposed to a generally increasing humoral immune response and SICCT or Bovigam® tests can give false negative results [162,165,166]. Therefore, late stage diseased animals are thought to be more accurately detected by serological tests, which are based on the detection of antibodies for *M. bovis* specific antigens [167]. Importantly, there is evidence that such skin test anergic animals are heavily diseased, shed higher amounts of bacteria and may be one cause of persistent and severe herd breakdowns [162,165,168]. Also, in countries without BTB control programs such animals are likely to be more prevalent and may contribute to a larger extent to the spread of the disease than animals at an early disease stage [165].

Some ELISA based serological tests for the diagnosis of BTB have been developed [162,169]. However, the currently available serological tests generally show a low sensitivity and could not yet replace the CMI response based diagnostic tests. More recently, a fluorescence polarization assay (FPA) for the detection of antibodies against MPB70 has been assessed for the diagnosis of BTB in naturally infected cattle [170]. Although test sensitivity was relatively low (62%), specificity was nearly 100% [170].

Vaccination against M. bovis infections

The development of a cattle vaccine against *M. bovis* infection has been considered a major priority for the control of BTB in the UK [171]. Provided an effective vaccine is available, cattle vaccination may be the most cost effective BTB control strategy and therefore also especially useful for interventions and disease control in developing countries [7]. The only currently available vaccine against tuberculosis in animals and humans is *M. bovis* BCG, which was developed through multiple passaging of a strain of *M. bovis* on glycerol soaked potato slices by Calmette and Guerin in 1921 [7]. However, BCG is not satisfactorily efficient in preventing disease and generally shows a variable efficacy. This variability has been attributed to several factors such as the vaccine strain itself [172], the regionally predominant infecting strains [173] or host genetics [7]. Moreover, it was shown that exposure to environmental NTM can influence the protective efficacy of BCG [174]. A particular obstacle of BCG is its sensitization of cattle to the tuberculin skin test and Bovigam®, making the discrimination of infected and vaccinated animals (differential diagnosis) impossible [155]. Current research in development of vaccines and new diagnostic tools must therefore be conducted in a coordinated manner to consider all these influencing factors in a pragmatic way [155].

Some recent studies could show an increased protective capacity or at least an improved CMI response upon administration of newly developed vaccine candidates compared to BCG [7]. A heterologous prime-boost strategy in cattle using a cocktail of three DNA vaccines for priming and BCG for boosting showed a better protection than BCG alone [175]. A similar approach using BCG for priming and boosting with modified vaccine virus Ankara (MVA) expressing the mycobacterial antigen Ag85A, showed very promising results in humans [176] and improved CMI responses in cattle [177].

Another strategy, which may offer increased protection, is the neonatal vaccination of calves. This approach has shown to enhance the Th1 bias of the immune response and circumvents the potential problems caused by pre-exposure to environmental mycobacteria, which may be of considerable importance in Africa [155,178].

Study rationale

BTB is primarily affecting poor populations in developing countries with no functional disease control programs. Moreover, very few information is available on the prevalence or distribution of BTB in these most severely affected regions of the world.

As shown before in the example of the UK, molecular genotyping of strains of *M. bovis* can help elucidate the general population structure of bacterial strains in a given area. Also, molecular typing of bacteria can be used to investigate the relationship of strains from different regions, to assess the spread of particular genotypes and to identify areas of strain-exchange.

In our study, we attempted to investigate the strain diversity and population structure of *M. bovis* in important livestock producing countries of different regions in Africa. The countries involved in our surveys were Senegal, Mali and Chad in the Sahel region of Africa, Tanzania as an important livestock producing country in East Africa and Algeria as a representative of the Maghreb region. By combining our data with the previously published records from other surveys in Africa and throughout the world, we were able to identify countries with identical genotypes of *M. bovis*. In particular, we were interested in identifying areas of strain exchange within Africa and evidence for the introduction of *M. bovis* from Europe into Africa, e.g. during the colonial period.

Our research has contributed in providing first insights into the distribution and relationship of *M. bovis* strains in different zones of Africa. These results are of great importance for the effective development of locally adapted interventions and control programs. Moreover, our work has contributed in deciphering the evolutionary history of *M. bovis* in Africa.

Attempts to evaluate diagnostic tests for BTB in naturally infected cattle in Africa are scarce but a prerequisite for the implementation of future surveillance schemes and control measures. BTB is characterized by a cell mediated immune response at early disease stages and a humoral immune response at later stages. Currently available diagnostic tests can be categorized on the basis of detecting either of both immune reactions.

Based on the hypothesis that in the absence of disease control schemes the frequency of late stage tuberculosis diseased cattle may be considerably higher in Africa compared to industrialized countries, we were interested in evaluating the intra-dermal comparative tuberculin skin test (a CMI response based test) and two recently developed fluorescence polarization assays (which are humoral response based diagnostic tests) in an African high BTB prevalence setting. The results of this study have important practical implications for BTB surveillance in the African context.

Aims and objectives

Aim

The aims of the study were (1) to contribute to the understanding of the population structure and evolutionary history of *Mycobacterium bovis* in Africa using molecular epidemiological techniques and (2) to evaluate multiple tests for the diagnosis of bovine tuberculosis in naturally infected cattle in Africa.

Objectives

Objective 1:

To assess the *M. bovis* strain diversities in Senegal, Mali, southern Chad, Tanzania and Algeria

Objective 2:

To investigate the interrelationship of *M. bovis* strain populations within Africa and between Africa and Europe

Objective 3:

To evaluate single intra-dermal comparative cervical tuberculin testing, two recently developed fluorescence polarization assays, meat inspection, microscopy, culture and PCR for the diagnosis of *M. bovis* infections in southern Chad

Objective 4:

To assess the importance of non-tuberculous mycobacteria and other infections for the diagnosis of bovine tuberculosis in southern Chad

Study sites and research partnerships

For our research activities in Africa we have established collaborations with institutions from our different study sites in Senegal, Mali, Chad, Tanzania and Algeria. The collaboration centers and the corresponding contact persons are indicated in figure 3 below.

For our work on the population structure of *Mycobacterium bovis* in Africa we have also set up a collaboration with Prof. R. Glyn Hewinson from the Veterinary Laboratories Agency (VLA), New Haw, Surrey, UK and Dr. Noel H. Smith from the University of Sussex, Falmer, UK.

In Switzerland, mycobacterial culture and part of the molecular work was performed in collaboration with Prof. E.C. Böttger at the Institute of Medical Microbiology (IMM) of the University Hospital in Zurich.

An integral part of the research collaboration with our African partners included the technology transfer of molecular epidemiological methods to several African laboratories. We have provided laboratory equipment to the TB research laboratories of the Ecole Inter-Etats des Sciences et Médecine Vétérinaires (EISMV) in Dakar, Senegal, Laboratoire de Recherches Vétérinaires et Zootechniques (LRVZ) in N'Djaména, Chad and the Sokoine University of Agriculture (SUA) in Morogoro, Tanzania. Laboratory technicians or students of the different institutes have been trained in PCR and VNTR typing. In particular, spoligotyping has been successfully (re-) established in the laboratory of Prof. R. Kazwala at SUA in Tanzania.

Dr. N. Sahraoui from the University of Saad Dahlab has been trained on PCR and VNTR typing at the Swiss Tropical Institute (STI) in Basel. Mr. B.N.R. Ngandolo has been trained on various molecular techniques at STI and VLA and he has received an extensive training on mycobacterial culture techniques at IMM.

Training of lab technicians has partially also been conducted under the umbrella of a Wellcome Trust funded network for BTB in Africa.

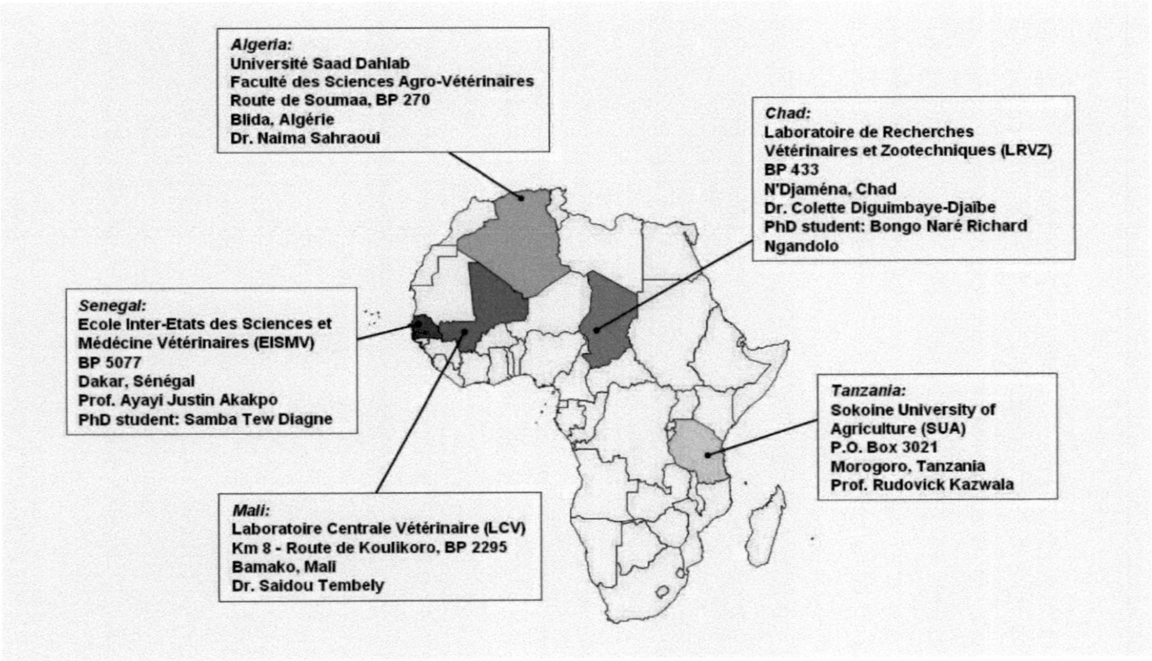

Algeria:
Université Saad Dahlab
Faculté des Sciences Agro-Vétérinaires
Route de Soumaa, BP 270
Blida, Algérie
Dr. Naima Sahraoui

Chad:
Laboratoire de Recherches
Vétérinaires et Zootechniques (LRVZ)
BP 433
N'Djaména, Chad
Dr. Colette Diguimbaye-Djaïbe
PhD student: Bongo Naré Richard Ngandolo

Senegal:
Ecole Inter-Etats des Sciences et Médécine Vétérinaires (EISMV)
BP 5077
Dakar, Sénégal
Prof. Ayayi Justin Akakpo
PhD student: Samba Tew Diagne

Tanzania:
Sokoine University of Agriculture (SUA)
P.O. Box 3021
Morogoro, Tanzania
Prof. Rudovick Kazwala

Mali:
Laboratoire Centrale Vétérinaire (LCV)
Km 8 - Route de Koulikoro, BP 2295
Bamako, Mali
Dr. Saidou Tembely

Figure 3: Overview of the collaborating institutions in Africa.

Part II

Molecular epidemiology of *Mycobacterium bovis* infections in Africa

Molecular characterization of *Mycobacterium bovis* isolated from cattle slaughtered at the Bamako abattoir in Mali

Borna Müller[1], Benjamin Steiner[1], Bassirou Bonfoh[2], Adama Fané[3], Noel H. Smith[4,5], Jakob Zinsstag[1]

[1]Swiss Tropical Institute, Basel, Switzerland
[2]Institut du Sahel, Bamako, Mali. Present address: Centre Suisse de Recherches Scientifiques en Côte d'Ivoire, Abidjan, Côte d'Ivoire
[3]Laboratoire Centrale Vétérinaire, Bamako, Mali
[4]Veterinary Laboratories Agency, Weybridge, United Kingdom
[5]University of Sussex, Falmer, United Kingdom

This article has been published in:
BMC Veterinary Research, 2008

Abstract

Background

Mali is one of the most important livestock producers of the Sahel region of Africa. A high frequency of bovine tuberculosis (BTB) has been reported but surveillance and control schemes are restricted to abattoir inspections only. The objective of this study was to conduct, for the first time, molecular characterization of *Mycobacterium bovis* strains isolated from cattle slaughtered at the Bamako abattoir. Of 3330 animals screened only 60 exhibited gross visible lesions. From these animals, twenty strains of *M. bovis* were isolated and characterized by spoligotyping.

Results

Organ lesions typical of BTB were most often detected in the liver, followed by the lung and the peritoneum. *M. bovis* was isolated from 20 animals and 7 different spoligotypes were observed among these 20 strains; three of the patterns had not been previously reported. Spoligotype patterns from thirteen of the strains lacked spacer 30, a characteristic common in strains of *M. bovis* found in Chad, Cameroon and Nigeria. However, unlike the other three Central African countries, the majority of spoligotype patterns observed in Mali also lacked spacer 6. Of the remaining seven strains, six had spoligotype patterns identical to strains commonly isolated in France and Spain.

Conclusions

Two groups of *M. bovis* were detected in cattle slaughtered at the Bamako abattoir. The spoligotype pattern of the first group has similarities to strains previously observed in Chad, Cameroon and Nigeria. The additional absence of spacer 6 in the majority of these strains suggests a Mali specific clone. The spoligotype patterns of the remaining strains suggest that they may have been of European origin.

Background

Bovine tuberculosis (BTB) is considered a neglected and poverty related zoonosis [26]. It has a major economic impact on livestock productivity [19], can persist in wildlife reservoirs and thus affect entire ecosystems [20] and it is of public health concern due to its zoonotic potential [5,23,32]. Although still prevalent in the developed world [132,154,179], BTB today mostly affects developing countries, which lack the financial and human resources to control the disease [5,32]. The Sahel region of Africa is extremely important in terms of animal production with Mali being amongst the principal cattle producing countries [180,181]. Mali has previously reported a high frequency of BTB but does not apply specific control measures, except carcass inspection at abattoirs [32]. In a recent prevalence study in dairy cattle herds from the peri-urban region of Bamako, 19% of the animals reacted positively to the comparative tuberculin skin test [46].

Spacer oligonucleotide typing (spoligotyping) [53] has been shown to be a valuable tool for investigations of the population structure of *Mycobacterium bovis* in a number of settings [43,52,61,132,179]. Furthermore, the international designation of spoligotype patterns (www.Mbovis.org, [136]) has facilitated the comparison of results from different countries and helps elucidate the distribution and spread of strains. Assuming that spoligotype spacers can only be lost and not regained, phylogenetic relationships between strains can be suggested [1,182]. A number of *Mycobacterium tuberculosis* complex (MTBC) strain families are readily identifiable through spoligotyping [135,183,184].

Variable number of tandem repeat (VNTR) typing is another simple method for *M. bovis* genotyping with a higher discriminatory power than spoligotyping [1,142]. However, extensive worldwide databases are presently not available and VNTR typing can today mainly be considered a valuable tool for sub-differentiation of strain groups initially identified by spoligotyping [1].

For *M. bovis*, previous studies in Chad, Cameroon and Nigeria have shown that virtually all spoligotype patterns lack spacer 30 with strains bearing spoligotype pattern SB0944 being the most frequent [43,52,61]. In Cameroon, because of the similarity to patterns of strains isolated in France [179] it was suggested that *M. bovis* could have been imported to this region during the French colonial period [61]. However, strains lacking spacer 30 were so far rarely found outside Chad, Cameroon and Nigeria.

The objectives of this study were to conduct an initial molecular characterization of *M. bovis* in Mali using spoligotyping and to identify potential exchange of strains with other regions.

Results

At the abattoir of Bamako, Mali, a case series of 3330 slaughter animals were sequentially screened during standard meat inspection in March and April 2007. A total of 182 specimens from 60 animals with gross visible lesions (apparent lesion prevalence: 1.8%; 95% CI: 1.4 – 2.3%) were collected. Organ lesions were most often detected in the liver (N = 22) followed by the lung (N = 14) and the peritoneum (N = 11). The specimens were put in culture and acid-fast bacilli were further characterized by spoligotyping and typing of the *M. bovis* specific RD4 region. Infection with *M. bovis* was confirmed for 20 animals. From two animals, strains of the *Mycobacterium fortuitum* complex were isolated as identified by partial sequencing of the 16S rRNA gene. In one case it appeared to be a single infection and in the other case a mixed infection of an *M. fortuitum* complex strain and *M. bovis*. In this animal, the *M. fortuitum* complex strain was isolated from liver lesions and *M. bovis* was isolated from lesions of the lungs and bronchial lymph nodes.

Infection with *M. bovis* was highly associated with the presence of lung lesions (N = 44, X^2 = 23.7, $p < 0.001$); in 79% of the animals exhibiting lung lesions, *M. bovis* infection could be confirmed. The association was less strong for liver lesions (N = 48, X^2 = 3.9, $p < 0.05$); only 41% of the animals with liver lesions were shown to be infected with *M. bovis*. However, in all 9 cases where *M. bovis* infection was detected in animals with liver lesions, lung lesions were present as well. No association was found between infection with *M. bovis* and lesions in organs other than the liver and lungs. Strains of *M. bovis* isolated in different organs of the same animal showed the same spoligotype pattern. Altogether among the 20 strains of *M. bovis* isolated, seven different spoligotypes were observed; four had been previously reported (SB0944, SB0300, SB0134 and SB0944) and the remaining three were designated SB1410, SB1411 and SB1412 by www.Mbovis.org (figure 4).

	SB number	Frequency	%	1	2	3	4	5	6	7	8	9	10	11	12	13	14	15	16	17	18	19	20	21	22	23	24	25	26	27	28	29	30	31	32	33	34	35	36	37	38	VNTR profile
	SB0944	1	5%	■	■		■	■	■	■	■		■	■	■	■	■	■		■	■	■	■	■	■	■	■	■	■	■	■	■		■	■	■	■	■	■	■	■	4 5 5 4* 3 3.1
	SB0300	3	15%	■	■		■	■		■	■		■	■	■	■	■	■		■	■	■	■	■	■	■	■	■	■	■	■	■		■	■	■	■	■	■	■	■	5 5 5 4* 3 3.1
	SB0300	3	15%	■	■		■	■		■	■		■	■	■	■	■	■		■	■	■	■	■	■	■	■	■	■	■	■	■		■	■	■	■	■	■	■	■	5 2 5 4* 3 3.1
	SB0300	1	5%	■	■		■	■		■	■		■	■	■	■	■	■		■	■	■	■	■	■	■	■	■	■	■	■	■		■	■	■	■	■	■	■	■	5 5 3/5 4* 3 3.1
Group I	SB0300	1	5%	■	■		■	■		■	■		■	■	■	■	■	■		■	■	■	■	■	■	■	■	■	■	■	■	■		■	■	■	■	■	■	■	■	5 5 3 4* 3 3.1
	SB1410	2	10%	■			■	■		■	■		■	■	■	■	■	■		■	■	■	■	■	■	■	■	■	■	■	■	■		■	■	■	■	■	■	■	■	5 6 5 4* 3 3.1
	SB1411	1	5%	■	■		■	■		■	■		■	■	■	■	■	■		■			■	■	■	■	■	■	■	■	■	■		■	■	■	■	■	■	■	■	7 4 5 4* 3 3.1
	SB1412	1	5%	■	■		■	■		■	■		■	■	■	■	■	■				■	■	■	■	■	■	■	■	■	■	■		■	■	■	■	■	■	■	■	3 4 5 4* 3 3.1
	Total	**13**	**65%**																																							
	SB0134	3	15%	■	■				■	■	■		■	■	■	■	■	■		■	■	■	■	■	■	■	■	■	■	■	■	■	■	■	■	■	■	■	■	■	■	7 5 5 4* 3 3.1
	SB0134	2	10%	■	■				■	■	■		■	■	■	■	■	■		■	■	■	■	■	■	■	■	■	■	■	■	■	■	■	■	■	■	■	■	■	■	6 5 5 4* 3 3.1
Group II	SB0134	1	5%	■	■				■	■	■		■	■	■	■	■	■		■	■	■	■	■	■	■	■	■	■	■	■	■	■	■	■	■	■	■	■	■	■	5 5 5 4* 3 3.1
	SB0991	1	5%	■	■				■	■	■		■	■	■	■	■	■		■	■	■	■	■	■	■	■	■		■	■	■	■	■	■	■	■	■	■	■	■	7 5 5 4* 3 3.1
	Total	**7**	**35%**																																							

(Columns 1–38: Spacer number)

Figure 4: Spoligotypes and VNTR typing patterns of *M. bovis* strains isolated from slaughter cattle at Bamako abattoir in Mali. Spacers 39-43 were absent from all spoligotype patterns. SB numbers were taken from www.Mbovis.org. VNTR typing targeted loci ETR A – F [138].

The distinctive lack of spacer 30 was observed in the spoligotype pattern of 13 strains; the majority of those in addition lacked spacer 6 (12/13; figure 4). The remaining seven strains were linked by the absence of spacers 4 and 5 (figure 4).

VNTR typing using the exact tandem repeats (ETR) A-F described by Frothingham et al. [138] allowed to further differentiate strains with the most frequent spoligotype patterns SB0300 and SB0134 (figure 4). Within the strains analyzed, only VNTR loci ETR A, B and C showed variation; ETR D, E and F profiles were identical in all the strains. One isolate exhibited two different VNTR alleles (3 and 5 tandem repeats) for locus ETR C (figure 4), indicating either a mixed infection with two distinct strains or a microevolution in this population of strains.

Discussion

The apparent prevalence of 1.8% gross visible lesions in Malian slaughter cattle was surprisingly low compared to published results from other Sahelian countries [43,63,64] and also lower than the previous tuberculin skin test results of cattle from Bamako indicated (reactor prevalence of 19%) [46]. This suggests that the cattle population slaughtered in the abattoir of Bamako originates largely from extensive pastoral rather than intensive peri-urban production systems, where BTB prevalence is usually higher [31]. Because lesion prevalence was so low and because we could not gauge the sensitivity of lesion detection we did not estimate the true prevalence of BTB at the Bamako abattoir. However, it is likely to be at least two to three fold higher than the observed prevalence [63,64].

Only the presence of lung lesions was strongly associated with the detection of *M. bovis* infection, suggesting that the lung was the primary site of *M. bovis* infection; the association of liver lesions and *M. bovis* infection was weaker. Furthermore, in all the cases where *M. bovis* infection could be confirmed in animals with liver lesions, lung lesions were also present (N = 9). In 12 animals liver lesions were recorded without associated lung lesions, however, in these animals we were unable to isolate *M. bovis* by culture. One animal exhibited a mixed infection of a *M. fortuitum* complex strain isolated from the liver and *M. bovis*, isolated from lesions associated with the lung and the bronchial lymph nodes. This indicates that some of the lesions outside the lung, and particularly the liver, have been caused by pathogens other than *M. bovis*. The amount of lesion causing infections due to other organisms than *M. bovis* might be considerable as only one third of the animals with lesions could be confirmed to be infected with *M. bovis*. Tuberculous gross visible lesions may be caused by an array of pathogens amongst which non-tuberculous mycobacteria (NTM) could play a crucial role in Africa (unpublished results, [58,69]). This suggestion is supported by a previous study in Chad where *M. fortuitum* complex was repeatedly isolated from lesions in slaughtered cattle [58]. It may be worthwhile to address the role of NTM infections on animal productivity in relation to BTB in African cattle.

Bacteria could only be isolated from 21 out of 60 animals with gross lesions; this was less than in other studies [43,69]. Failure to cultivate bacteria could have been due to long-term storage of tissue samples in the freezer prior to cultivation. Because of frequent power cuts, we cannot exclude that some of the specimen might have undergone multiple freeze-thaw cycles while they were stored. A high amount of completely calcified lesions, without viable tubercle bacilli, could also explain the low recovery of bacteria (references within [64]).

The strains identified here can be divided into two groups by spoligotype pattern (figure 4). The first group is marked by the distinct loss of spacer 30 and accounts for 65% of the strains detected (figure 4). The absence of spacer 30 has previously found to be characteristic of spoligotype patterns for strains of *M. bovis* isolated in Chad, Cameroon and Nigeria [43,52,61]. In addition to the absence of spacer 30, most of the strains from Mali (12/13) also lack spacer 6, a characteristic not seen in isolates from Central African countries [43,52,61]. The most often detected *M. bovis* strains in Mali with spoligotype pattern SB0300 could have evolved from strains with spoligotype pattern SB0944 either by drift or a selective sweep. This is supported by the fact that two of the three VNTR types identified in Malian strains with spoligotype pattern SB0300 were identical to VNTR types of previously isolated Nigerian *M. bovis* strains with spoligotype pattern SB0944 [52]. Altogether, the results suggest a close relationship between strains from Mali and those from Central Africa. The spread of related strains over this large area could be explained by the predominant long distance transhumant livestock production system in the Sahel, mainly practiced by Fulbe pastoralists [185]. However, considering the fact that, except for pattern SB0944, none of the spoligotype patterns found in Mali are present in any of the three Central African countries and considering that strains with spoligotype pattern SB0944 were rarely detected in Mali we suppose that the spread of *M. bovis* strains over this large distance is relatively slow.

The second group of related spoligotype patterns is characterized by the absence of spacers 4 and 5 (figure 4). The most often detected spoligotype belonging to this group is commonly found in strains from France and Spain (SB0134 [179,186]), suggesting a link between *M. bovis* strains from Mali and mainland Europe. VNTR profile 6 5 5 4 for ETR loci A-D identified in 2/6 Malian strains of *M. bovis* with spoligotype pattern SB0134 has also been detected in a strain isolated from French cattle in the Normandy in 1996 with the same spoligotype pattern [187]. Moreover, three other Malian *M. bovis* strains with ETR A-D profile 7 5 5 4 and spoligotype pattern SB0134 could be closely related to SB0134 *M. bovis* strains with ETR A-D profile 7 4 5 4, which is frequently found in *M. bovis* strains from the Normandy [187]. However, identical spoligotype patterns have also been found in strains from northern Algeria (unpublished results) and livestock migrations from Algeria to Mali through the Sahara desert have been reported. Comprehensive genotyping of strains from West Africa, North Africa and Europe using highly polymorphic markers would be necessary to further elucidate the interrelationship of *M. bovis* strains from these different regions.

Njanpop-Lafourcade et al., have previously suggested an influence of the French colonial history based on the similarity of the predominant spoligotype pattern (SB0944) in Cameroon to the BCG-like spoligotype pattern that is commonly seen in strains from France [61]. In a similar manner it is possible to suggest that strains of *M. bovis* with spoligotype pattern SB0134 were originally imported from Europe. If both assumptions are true this would suggest that either *M. bovis* was not present in the Central or West African region before introduction from Europe to Africa or previously existing "native" *M. bovis* strains have been largely replaced.

Due to the small sample size, the limited survey period and the sampling at only one study location, the *M. bovis* strains collected cannot reflect the country-wide bacterial population structure. Therefore, frequencies of strains with a specific spoligotype may not necessarily mirror the actual frequency of these strains in the population although more frequent strains are also more likely to be detected in a random sample. Moreover, other groups of *M. bovis* strains than the two that were observed may be present in Mali. However, due to the predominant long distance transhumant livestock production system, we believe that the slaughter cattle encountered at the abattoir of Bamako and consequently their associated *M. bovis* strains, represent a sample from a large area of the country.

Conclusions

This study presents the first molecular characterization of *M. bovis* strains from Mali. The results suggest that the most often detected strains are related to strains that are prevalent in Chad, Cameroon and Nigeria. A second group of strains shows spoligotype patterns similar to those abundant in mainland Europe and could have been imported directly from Europe or via northern Africa. Our results can serve as a baseline study for future comparisons with strains from other areas in and around Mali.

Methods

Sample collection

Samples were collected from a sequential series of slaughter cattle at the Bamako abattoir, Mali in March and April 2007. The cattle population consisted of crossbreeds between N'Dama, zebu and exotic breeds. The origin of the cattle could not be traced due to poor documentation and multiple selling-on of the animals before slaughter. However, we originally assumed that the cattle originate from the peri-urban region of Bamako as well as the principle areas of cattle production throughout the country. After slaughter, animals underwent a standard meat inspection and organs showing gross visible lesions were confiscated. No ethical clearance was required for this study because it was done on slaughtered animals and organ confiscation was part of a routine monitoring. Tissue samples of 60 animals with gross visible lesions were collected. The samples were transported on ice to the Central Veterinary Laboratory in Bamako. Upon arrival, samples were immediately liberated from connective tissue and fat under a bio-safety cabinet and by use of sterile dissection instruments. The samples were seared on the outside in order to reduce superficial contamination, sealed into sterile stomacher bags and stored at -20°C for maximum one and a half months until they were shipped to Switzerland for culture. Because of frequent power cuts we cannot exclude that some of the specimen might have undergone multiple freeze-thaw cycles while they were stored. Shipment to Switzerland occurred in a refrigerated box; the temperature of the samples was monitored at all times by use of a data-logger (HoboTemp by OnsetCorp) and never exceeded -10°C.

Tissue preparation, culture and DNA extraction

At the Swiss Reference Centre for Mycobacteria, samples were stored at -80°C until processed. Specimen were dissected and approximately 2 g were homogenized for 2 minutes in 10 ml phosphate buffer saline (PBS) using the ULTRA-TURRAX® Tube Drive homogenizer with DT-20 tubes (IKA, Staufen, Germany). A 5 ml aliquot of the suspension was decontaminated for 15 minutes with 5 ml of decontamination solution (0.5% N-acetyl-L-cystein/2% NaOH/1.45% Na-Citrate solution). The decontamination was stopped by addition of 15 ml PBS, the suspension was centrifuged at 3500 rpm for 15 minutes and the pellet was re-suspended in 2 ml PBS. Then, 0.5 ml of the suspension was added to a BBL™ MGIT™ Mycobacterium Growth Indicator Tube containing OADC enrichment and PANTA™ antibiotic mixture (BD) and incubated in a BACTEC™ MGIT™ 960 mycobacterial Detection System. In addition, 0.25 ml of the suspension was inoculated onto Löwenstein-Jensen and 7H10 culture media and incubated at 37°C. Cultures were incubated until growth was detected or for at least 8 weeks. Presence of acid-fast bacilli was tested by Ziehl-Neelsen staining and Microscopy and DNA of positive cultures was extracted using the InstaGene™ Matrix (Bio-Rad®).

Molecular characterization

Spoligotyping was conducted as previously described [53]. VNTR typing was performed according to the method of Frothingham et al. [138] with adaptations described elsewhere [86]. According to the findings of Brosch et al., strains were confirmed as *M. bovis* by the absence of region RD4. [145]. The 16S rRNA gene amplification and sequencing was carried out as described by Zucol et al., (2006) [188]. Species identification was carried out by comparison with the sequences of the SmartGene Integrated Database Network System (IDNS^{TM}) 3.4.0.

Statistical analysis

Statistical analysis was carried out using Intercooled Stata 9.2 for Windows (StataCorp LP, USA). The association between the presence of lesions in each organ and confirmed *M. bovis* infection was tested by a chi-squared test.

Acknowledgements

We would like to thank Prof. Erik C. Böttger and the technicians of the Swiss National Centre for Mycobacteria in Zurich for assistance and laboratory facilities for culturing of mycobacteria. We would also like to thank Prof. R. Glyn Hewinson and Dr. Stefan Berg from the Veterinary Laboratories Agency (VLA) in Weybridge, UK for providing technical support and laboratory facilities for spoligotyping and K. Gover for technical assistance. Moreover, we are indebted to the director of the Laboratoire Centrale Vétérinaire in Bamako and the meat inspectors at the abattoirs for providing laboratory facilities and assistance during the sample collection. Our work has received financial support from the Swiss National Science Foundation (project no. 107559).

Molecular characterization of *Mycobacterium bovis* strains isolated from cattle slaughtered at two abattoirs in Algeria

Naima Sahraoui[1,2]*, Borna Müller[3]*, Djamel Guetarni[1], Fadéla Boulahbal[4], Djamel Yala[4], Rachid Ouzrout[2], Stefan Berg[5], Noel H. Smith[5,6], Jakob Zinsstag[3]

[1]Université Saad Dahlab, Blida, Algeria
[2]Centre Universitaire d'El-Tarf, El-Tarf, Algeria
[3]Swiss Tropical Institute, Basel, Switzerland
[4]Institut Pasteur d'Algérie, Algiers, Algeria
[5]Veterinary Laboratories Agency, Weybridge, United Kingdom
[6]University of Sussex, Falmer, United Kingdom

*These authors contributed equally to this work

This article has been published in:
BMC Veterinary Research, 2009

Abstract

Background

Bovine tuberculosis is prevalent in Algeria despite governmental attempts to control the disease. The objective of this study was to conduct, for the first time, molecular characterization of a population sample of *Mycobacterium bovis* strains isolated from slaughter cattle in Algeria. Between August and November 2007, 7250 animals were consecutively screened at the abattoirs of Algiers and Blida. In 260 animals, gross visible granulomatous lesions were detected and put into culture. Bacterial isolates were subsequently analyzed by molecular methods.

Results

Altogether, 101 bacterial strains from 100 animals were subjected to molecular characterization. *M. bovis* was isolated from 88 animals. Other bacteria isolated included one strain of *M. caprae*, four *Rhodococcus equi* strains, three non-tuberculous mycobacteria (NTM) and five strains of other bacterial species. The *M. bovis* strains isolated showed 22 different spoligotype patterns; four of them had not been previously reported. The majority of *M. bovis* strains (89%) showed spoligotype patterns that were previously observed in strains from European cattle. Variable number of tandem repeat (VNTR) typing supported a link between *M. bovis* strains from Algeria and France. One spoligotype pattern has also been shown to be frequent in *M. bovis* strains from Mali although the VNTR pattern of the Algerian strains differed from the Malian strains.

Conclusions

M. bovis infections account for a high amount of granulomatous lesions detected in Algerian slaughter cattle during standard meat inspection at Algiers and Blida abattoir. Molecular typing results suggested a link between Algerian and European strains of *M. bovis*.

Background

Mycobacterium bovis is the causative agent of bovine tuberculosis (BTB) and belongs to the *Mycobacterium tuberculosis* Complex (MTBC), a group of closely related bacteria causing tuberculosis in various mammalian hosts [1].

BTB has a major economic impact on livestock productivity [19], can persist in wildlife and thus affect entire ecosystems [20] and it is of public health concern due to its zoonotic potential [5,23,32]. Although still prevalent in the developed world [132,154,179], BTB today mostly affects developing countries, which lack the financial and human resources to control the disease [5,32]. BTB is also known to be prevalent in Algeria despite governmental attempts to control the disease [5,32]. However, control is restricted to abattoir meat inspection and biannual intra-dermal tuberculin skin testing of cattle from intensive dairy farms [189]. Moreover, the majority of Algerian cattle are not registered and cattle movement control schemes are not well established. Most of the BTB cases in Algeria are discovered during meat inspection in slaughter cattle at abattoirs when gross visible lesions typical of the disease are detected. However, in two recent studies in Chad and Uganda, non-tuberculous mycobacteria (NTM) were isolated from more than 40% of the animals exhibiting lesions [58,69]. This suggests that NTM infections in cattle might be of considerable importance in some African countries.

Spacer oligonucleotide typing (spoligotyping) and variable number of tandem repeat (VNTR) typing have been shown to be valuable tools for molecular epidemiology of *M. bovis* infections in a number of settings [61,138,142,179]. Spoligotyping can be used to identify distinct groups of strains, which can often be further differentiated by VNTR typing due to the latter's higher discriminatory power [1]. Extensive worldwide databases of spoligo- and VNTR typing patterns facilitate the comparison of results from different countries and help to elucidate the distribution and spread of strains (www.Mbovis.org, [136]). The objectives of this study were to molecularly characterize a population sample of strains of *M. bovis* from Algeria using spoligotyping and VNTR typing and to identify potential exchange of strains with other regions.

Results

At the abattoirs of Algiers and Blida in Algeria a consecutive case series of altogether 7250 slaughter animals was examined during standard meat inspection from August to November 2007. Lesions suggestive of BTB were sampled from 260 animals (apparent lesion prevalence: 3.6%; CI: 3.2 – 4.0%) and put into culture. Cultures from 106 animals with lesions did not show bacterial growth. Cultures from 20 animals showed contaminations and the remaining cultures from altogether 134 animals showed bacterial growth without visible contaminations. A sample of bacterial cultures from altogether 100 of these animals were further characterised by deletion-, spoligo- and VNTR typing, and sequencing of the 16S rRNA gene. Strains of *M. bovis* were identified by the deletion of RD4 in the genome sequence and by spoligotyping; *M. bovis* was detected in samples from 88 animals (table 2). In cultures of one of these animals in addition to *M. bovis*, a strain of *Rhodococcus equi* could be detected. One strain of the *M. caprae* clade was identified by the presence of the RD4 region and the deletion of RD12 in its genome sequence [1]. Altogether, four animals were shown to be infected with *R. equi* (including the animal with the *M. bovis/R. equi* mixed infection). NTM infections were detected in three animals (table 2). The 16S rRNA gene sequence of one NTM strain was most similar to the 16S rRNA gene sequence of *M. chitae* (97.7% sequence identity); the second NTM strain was most closely related to *M. brasiliensis* (98.4% sequence identity) and the third NTM strain showed highest sequence similarities to *M. acapulcensis* and *M. flavescens* (99.7% sequence identity for both). However, species identification of the NTM strains by 16S rRNA sequencing did not meet the requirements reported by Bosshard et al. and may have to be considered with caution [190]. *Ureibacillus thermosphaericus* and bacteria of the genus *Corynebacterium*, *Paenibacillus*, *Pseudomonas* and *Streptococcus* were each detected in samples from one animal (table 2).

	N	%
Animals screened	**7250**	**100%**
Animals with lesions	**260**	**4%**
Bacterial strains isolated:	**101**	**100%**
*Mycobacterium bovis**	88	87%
Mycobacterium caprae	1	1%
Non-tuberculous Mycobacteria	3	3%
*Rhodococcus equi**	4	4%
Ureibacillus thermosphaericus	1	1%
Corynebacterium spp.	1	1%
Paenibacillus spp.	1	1%
Pseudomonas spp.	1	1%
Streptococcus spp.	1	1%

* One mixed infection of *M. bovis* and *R. equi* was detected in one animal

Table 2. Identification of 101 bacteria isolated from tuberculous lesions of 100 slaughtered cattle in Algeria.

Spoligotyping of the 89 MTBC strains isolated revealed altogether 23 different spoligotype patterns (figure 5); 18 of them had been previously reported and 5 were new, including the pattern of the *M. caprae* strain [146,191,192]. Previously unreported *M. bovis* spoligotype patterns were named SB1447, SB1448, SB1449 and SB1450 by www.Mbovis.org and the new *M. caprae* spoligotype pattern was named SB1451 (figure 5). The four most frequent spoligotype patterns (SB0120, SB0121, SB0134, SB0941) accounted for 40%, 22%, 7% and 7% of the *M. bovis* strains, respectively. Of the 22 *M. bovis* spoligotypes, 8 were clustered and the remaining 14 were unique patterns (allelic diversity = 0.78).

All the MTBC strains were VNTR typed using the loci ETR A-E (see appendix 2) [138]. For the *M. bovis* strains, 35 VNTR types with 16 clustered and 19 unique patterns could be identified (allelic diversity = 0.93). Spoligotyping combined with VNTR typing allowed us to distinguish 51 distinct types of *M. bovis* strains with 11 of them being clustered (allelic diversity = 0.95). The VNTR patterns of the four most common spoligotypes (SB0120, SB0121, SB0134 and SB0941) are shown in table 3. Spoligo- and VNTR types of all the 89 MTBC strains are summarized in appendix 2. It is noteworthy that the frequency of the various *M. bovis* genotypes detected did not markedly differ between the two study locations Algiers and Blida.

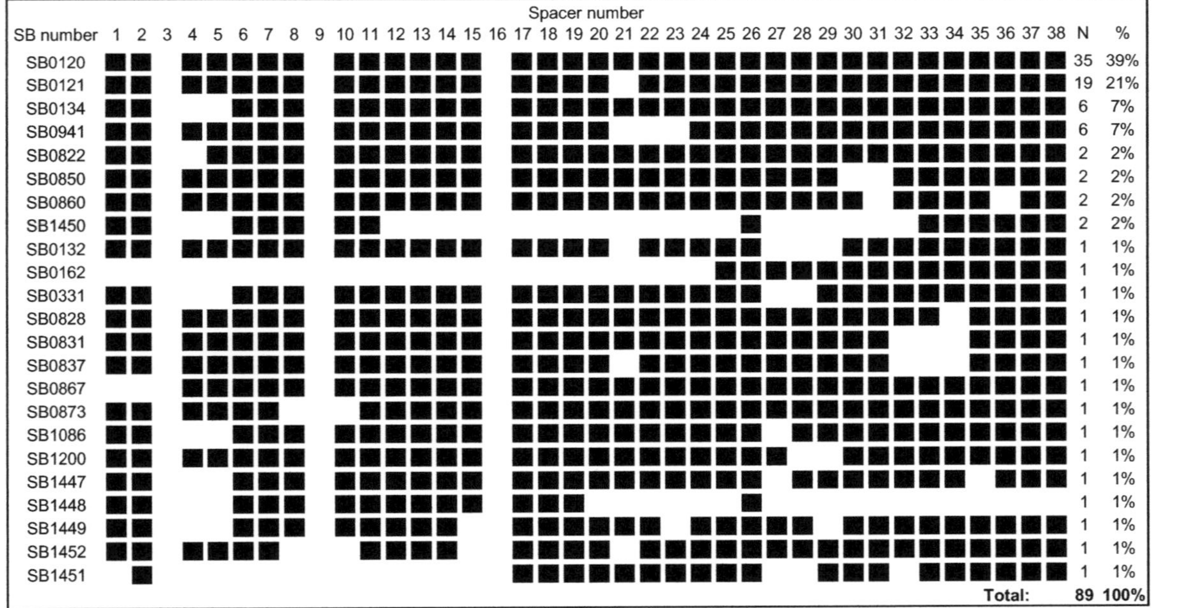

Figure 5. Spoligotype patterns of MTBC strains isolated from slaughter cattle at the abattoirs of Algiers and Blida in Algeria. Spacers 39-43 were absent from all spoligotype patterns. SB numbers were taken from www.Mbovis.org. Previously unreported *M. bovis* spoligotype patterns were named SB1447, SB1448, SB1449 and SB1450 by www.Mbovis.org and the new *M. caprae* spoligotype pattern was named SB1451.

SB0120		SB0121		SB0134		SB0941	
VNTR type	Frequency	VNTR type	Frequency	VNTR type	Frequency	VNTR type	Frequency
4 5 5 4* 3	6	6 4 3 4* 3	16	7 4 5 4* 3	3	6 5 3 4* 3	5
5 5 2 4* 3	4	4 3 5 3* 3	1	3 5 5 5* 3	1	6 7 3 4* 3	1
4 5 3 4* 3	3	5 3 5 4* 3	1	4 4 5 3* 4	1		
4 5 5 3* 3	3	6 4 4 4* 3	1	7 5 3 5* 3	1		
4 4 5 4* 3	2						
4 5 5 2* 3	2						
5 3 5 4* 3	2						
5 5 5 4* 4	2						
3 2 5 4* 3	1						
3 4 5 3* 3	1						
3 5 5 4* 4	1						
4 5 2 4* 3	1						
4 7 3 4* 3	1						
4 8 5 4* 3	1						
5 4 4 4* 3	1						
5 4 5 4* 3	1						
5 5 5 4* 3	1						
5 6 5 4* 3	1						
7 5 4 3* 3	1						
Total	35		19		6		6

Table 3. VNTR allele profiles (in order of loci ETR A-E) of strains with spoligotype patterns SB0120, SB0121, SB0134 and SB0941.

Discussion

To our knowledge, this is the first study conducting the molecular characterization of strains of *M. bovis* isolated from slaughter cattle in Algeria. However, routine abattoir meat inspection and periodic intra-dermal tuberculin skin testing of cattle from intensive dairy farms have already previously revealed the presence of BTB in Algeria [5,32]. In our study, MTBC infections were detected in 89/7250 animals (1% of all sampled animals). Considering the imperfect sensitivity of meat inspection and culture and the fact that only 100 of the 134 positive cultures have been characterized, the true prevalence of MTBC infections in the sampled cattle population in fact may has been considerably higher. Ayele et al. previously reported a low sporadic incidence of BTB in Algeria [15]; therefore, our results indicate a more frequent occurrence of *M. bovis* infections in Algerian cattle than previously suspected.

Unlike reported for Chad and Uganda, where NTM strains were isolated from more than 40% of the animals with lesions [58,69], only 3 of 100 Algerian cattle showed NTM infections. *R. equi*, which was detected in samples of 4 animals, has so far not been reported in connection with cattle infections in Africa. Taken together, these results suggest that there may be a difference in the bacterial species causing pulmonary infection and lesion formation in cattle between Algeria and some countries of sub-Saharan Africa.

The calculated allelic diversity of spoligotype patterns among the *M. bovis* strains isolated in Algeria (0.78) was relatively low compared to France (0.91) but similar to strain diversities reported from the UK (0.79) and many other places in the world [1,43,61,179]. However, considering the fact that the sample of *M. bovis* genotypes characterized in our study may not be representative of the country wide population of *M. bovis*, the calculated allelic diversity should be considered with care.

Of the 22 *M. bovis* spoligotype patterns detected, 13 patterns, accounting for 89% of all the strains isolated, have previously been detected in strains isolated from French cattle [179]. The three spoligotype patterns most frequently detected in strains from Algeria (SB0120, SB0121, SB0134) are also the three most frequent types observed in France [179] and are also known to be common in strains from other continental European countries [97,131,132,193]. Live animal importation from Europe to Algeria is documented. It presumably started during the French colonial period (1830-1962), has continued up to the present time and has mainly aimed at increasing the Algerian population of highly productive dairy cattle. Therefore, the observed similarity of spoligotypes may reflect the introduction of *M. bovis* from mainland Europe to Algeria.

Interestingly, in a previous study we have identified strains of spoligotype pattern SB0134 (figure 5) as highly prevalent in cattle from neighbouring Mali [47]. Strains of that spoligotype pattern are also the third most frequent strains detected in France. VNTR typing results for loci ETR A-E were available for strains from Algeria and Mali but no matches were found (table 4). ETR A-D patterns for strains with spoligotype pattern SB0134 from the Normandy region in France have previously been published [187]. Pattern 7 4 5 4 for loci ETR A-D was detected in strains from France and Algeria and pattern 6 5 5 4 for loci ETR A-D was present in France and Mali (table 4). We obtained VNTR patterns of *M. bovis* strains isolated in France with spoligotype pattern SB0120 (M. Boschiroli, unpublished data) and found matching ETR A-D patterns for 10/19 Algerian *M. bovis* strains with spoligotype pattern SB0120 (data not shown).

Algeria	Mali	France
3 5 5 5* 3	5 5 5 4* 3	2 4 5 4
4 4 5 3* 4	6 5 5 4* 3	3 3 3 4
7 4 5 4* 3	7 5 5 4* 3	3 4 5 7
7 5 3 5* 3		4 5 5 4
		6 5 5 4
		7 4 5 4

Table 4. ETR A-E and ETR A-D typing results of strains with spoligotype pattern SB0134 from Algeria, Mali [47] and from the region of Normandy in France [187], respectively. Matching patterns of strains are indicated in color. The presence of a 24 bp deletion in one of the repeats of ETR D (in our analysis marked with *) was not declared in the French study.

Due to the small sample size in the studies in Mali and Algeria and the sampling at only two abattoirs in northern Algeria and one abattoir in Mali, we cannot infer the absence of strain exchange between Algeria and neighbouring Mali. However, our data indicates that some strains of *M. bovis* found in Algeria and

Mali may have been independently introduced from France (or more generally continental Europe). Live animal importation from Europe into Algeria is continuing up to date. However, nowadays, a negative tuberculin skin testing result must be certified before importation. During the colonial period, import restrictions may have been less rigorous for cattle imported from France. Also, the increased efforts to control BTB in Europe during the second half of the 20th century have lead to a decreased prevalence of the disease in cattle of many European countries [194]. Therefore, supposedly, introduction of strains of *M. bovis* into Algeria could have most likely occurred during the colonial time. However, occasional importation of diseased but undetected live cattle might still be possible. Introduction of *M. bovis* from Europe into other countries has been suggested several times [61,134,179,195,196] and further investigations on the relationship between European strains and strains from other parts of the world might be worthwhile in order to elucidate the global spread of *M. bovis*.

Due to the small sample size, the limited survey period and the sampling at only two abattoirs in northern Algeria, the population sample of *M. bovis* strains collected may not reflect the country-wide bacterial population structure. Therefore, frequencies of strains with a specific spoligotype do not necessarily mirror the actual frequency of these strains in the population. Moreover, it is possible that animals from some regions have been overrepresented in our sample. The origin of the animals could not be traced due to poor documentation and multiple selling-on of the animals prior to slaughter. However, we assume that the majority of animals were dairy cattle and that only few animals originated from the same herds. Due to the characteristics of the local livestock production system we further assume that the animals originated from a large area of northern Algeria with a majority of them presumably coming from the central northern part. Indeed, there was evidence that some of the cattle slaughtered at the abattoir of Algiers originated from an approximately 300km distant region around Sétif. The implementation of a tracing system for animals would be of great value to further enhance BTB surveillance through abattoir meat inspection or molecular epidemiological studies. A tracing system could possibly also help in the early detection of BTB outbreaks and their localization.

Conclusions

This study presents the first molecular characterisation of a population sample of strains of *M. bovis* isolated from Algerian cattle. BTB accounted for a high amount of granulomatous lesions detected in Algerian slaughter cattle during standard meat inspection at Algiers and Blida abattoir. Spoligotyping as well as VNTR typing results suggest a close link between the strains isolated from Algerian cattle and *M. bovis* strains from mainland Europe. This study highlights the importance of both spoligotype and VNTR typing databases and standardized protocols to assist global molecular epidemiological investigations of *M. bovis* infections.

Methods

Sample collection

Samples were collected between August and November 2007 from a sequential series of slaughter cattle at two abattoirs in Algeria (in Algiers and Blida), approximately 50 km apart from each other. The cattle population consisted mainly of young males and old cows with males being significantly more often slaughtered at the abattoir of Algiers and cows more often slaughtered at the abattoir of Blida. Of altogether 7250 animals examined, 93% of the cattle were exotic breeds (Holstein and Montbelliard), 6% were cross-breeds and only 1% local breeds. Altogether, 4980 animals (69%) were males and 2270 (31%) were females. The origin of the cattle could not be traced due to poor documentation. We assume that the majority of animals were dairy cattle and only few animals from the same herds. The animals encountered at Algiers and Blida abattoir possibly originated from a large area of northern Algeria with a majority of them presumably coming from the central northern part. Tissue samples of 260 animals with gross visible lesions were collected. The samples were transported on ice to the Institut Pasteur d'Alger for further processing.

Tissue preparation, culture and DNA extraction

At the Institut Pasteur in Algiers, specimens from all 260 animals, which exhibited gross visible lesions, were dissected and manually homogenised using a mortar. Samples were decontaminated by addition of 4ml of 4% H_2SO_4

and neutralised with 6% NaOH using bromothymol blue as an indicator for the pH. Two Löwenstein-Jensen slants, supplemented with either sodium pyruvate or glycerol, were inoculated with 3 ml of the suspension and incubated at 37°C until bacterial growth was visible or for at least 12 weeks. Presence of Acid-Fast Bacilli was tested by Ziehl-Neelsen staining and microscopy. Cultures from 106 animals did not show bacterial growth. Cultures from 20 animals showed contaminations and the remaining cultures from altogether 134 animals showed bacterial growth without visible contaminations.

A sub-sample of 101 bacterial cultures from 100 animals was sent to the National Reference Centre for Mycobacteria in Zurich, Switzerland. DNA of all cultures was extracted using the InstGene™ Matrix (Bio-Rad®).

Molecular characterization

Spoligotyping was conducted at the Veterinary Laboratories Agency in Weybridge, UK as previously described [53]. VNTR typing was carried out as previously described using primers targeting the loci ETR A, ETR B, ETR C, ETR D and ETR E according to protocols of Frothingham et al. [138]. The ETR-D locus contains a 24 bp

deletion in one of the repeats and the naming convention indicates the presence of this deletion by a * i.e. 4* (= 3 x 77 bp repeats and one 53 bp repeat) [86]. Allelic diversity was calculated according to the method of Selander et al. [197]. Strains were identified as *M. bovis* by the absence of the region RD4 and as *M. caprae* by the absence of RD12 and presence of RD4 as described by Brosch et al. [145]. The 16S rRNA gene amplification and sequencing was carried out as described by Zucol et al. [198]. Species identification was carried out by comparison with the sequences of the SmartGene Integrated Database Network System (IDNS™) 3.4.0. Criteria for species identification were taken from Bosshard et al. [190].

Acknowledgements

We would like to thank Prof. Erik C. Böttger and the technicians of the Swiss National Centre for Mycobacteria in Zurich for assistance and laboratory facilities. We would also like to thank Prof. R. Glyn Hewinson from the Veterinary Laboratories Agency in Weybridge, UK for providing technical support and laboratory facilities for spoligotyping and VNTR typing. Dr. Maria L. Boschiroli has kindly provided us with preliminary VNTR typing data of *M. bovis* strains isolated from French cattle. Our work has received financial support from the Swiss National Science Foundation (project no. 107559) and the Wellcome Trust Livestock for Life initiative.

African 1; an epidemiologically important clonal complex of *Mycobacterium bovis* dominant in Mali, Nigeria, Cameroon and Chad

Borna Müller[1*], Markus Hilty[2*], Stefan Berg[3*], M. Carmen Garcia-Pelayo[3], James Dale[3], Laura Boschiroli[4], Simeon Cadmus[5], Bongo Naré Richard Ngandolo[6], Colette Diguimbaye-Djaibé[6], Rudovick Kazwala[7], Bassirou Bonfoh[8], B.M. Njanpop-Lafourcade[9], Naima Sahraoui[10,11], Djamel Guetarni[10], Abraham Aseffa[12], Meseret H. Mekonnen[12], Voahangy Rasolofo Razanamparany[13], Herimanana Ramarokoto[13], Berit Djønne[14], James Oloya[15], A. Machado[16], Custodia Mucavele[16], Eystein Skjerve[17], Francoise Portaels[18], Leen Rigouts[18], Anita Michel[19], Annélle Müller[20], Gunilla Källenius[21], Paul D. van Helden[20], R. Glyn Hewinson[3], Jakob Zinsstag[1], Stephen V. Gordon[22] and Noel H. Smith[3,23]

*These authors contributed equally to this manuscript

[1]Swiss Tropical Institute, Basel, Switzerland

[2]Imperial College London, London, United Kingdom

[3]Veterinary Laboratories Agency, Weybridge, United Kingdom

[4]Agence française de sécurité sanitaire des aliments, Maisons-Alfort, France

[5]University of Ibadan, Ibadan, Nigeria

[6]Laboratoire de Recherches Vétérinaires et Zootechniques, N'Djaména, Chad

[7]Sokoine University of Agriculture, Morogoro, Tanzania

[8]Centre Suisse de Recherches Scientifiques, Abidjan, Côte d'Ivoire

[9]Institut Pasteur, Paris, France

[10]Université Saad Dahlab, Blida, Algeria

[11]Centre Universitaire d'El-Tarf, El-Tarf, Algeria

[12]Armauer Hansen Research Institute, Addis Ababa, Ethiopia

[13]Institut Pasteur de Madagascar, Antananarivo, Madagascar

[14]National Veterinary Institute, Oslo, Norway

[15]Makerere University, Kampala, Uganda

[16]Universidade Eduardo Mondlane, Maputo, Mozambique

[17]Norwegian School of Veterinary Science, Oslo, Norway

[18]Institute of Tropical Medicine, Antwerp, Belgium

[19]ARC-Onderstepoort Veterinary Institute, Onderstepoort, South Africa

[20]Stellenbosch University, Tygerberg, South Africa

[21]Karolinska Institutet, Stockholm, Sweden

[22]University College Dublin, Dublin, Ireland

[23]University of Sussex, Brighton, UK

This article has been published in:

Journal of Bacteriology, 2009

Abstract

We have identified a clonal complex of *Mycobacterium bovis* present at high frequency in cattle in population samples from several sub-Saharan, West-Central African countries. This closely related group of bacteria is defined by a specific chromosomal deletion (RDAf1) and can be identified by the absence of spacer 30 in the standard spoligotype typing scheme. We have named this group of strains the African1 (Af1) clonal complex and defined the spoligotype signature of this clonal complex as the same as BCG vaccine strain but with the deletion of spacer 30.

Strains of the Af1 clonal complex were found at high frequency in population samples of *M. bovis* from cattle in Mali, Cameroon, Nigeria and Chad and using a combination of VNTR typing and spoligotyping we show that the population of *M. bovis* in each of these countries is distinct, suggesting that the recent mixing of strains between countries is not common in this area of Africa.

Strains with the Af1 specific deletion (RDAf1) were not identified in *M. bovis* isolates from Algeria, Burundi, Ethiopia, Madagascar, Mozambique, South Africa, Tanzania and Uganda. Furthermore, the spoligotype signature of the Af1 clonal complex has not been identified in population samples of bovine TB from Europe, Iran and South America. These observations suggest that the Af1 clonal complex is geographically localized, albeit to several African countries, and we discuss the demographic and evolutionary scenarios that may have led to this localization.

Author Summary

This article summarizes the work of a very large collaboration of African and European scientists working to understand the origin, distribution and evolution of bovine tuberculosis (*Mycobacterium bovis*) in cattle in Africa. We identify a closely related clonal complex of *M. bovis* strains that are present in high frequency in Mali, Cameroon, Chad and Nigeria. The strains of this clonal complex all possess the same chromosomal deletion and therefore, in this clonal organism, are all descended from the same ancestral cell. We show that this clonal complex of strains, called African 1 (Af1), is geographically localized to these sub-Saharan, West-Central African countries and is not prevalent in East African countries, Europe or South America.

Surprisingly, by advanced genotyping of the strains, it was possible to identify country specific populations of Af1 suggesting that the recent movement of bovine TB throughout this region of Africa is not common. This manuscript shows how the identification and analysis of bovine TB clonal complexes can help to elucidate the demography and evolution of this important veterinary and human pathogen as well as pointing the way to future work that may generate important epidemiological methods to help in the identification and elimination of the disease.

Introduction

Mycobacterium bovis causes bovine tuberculosis (bovine TB), an important disease of domesticated cattle that has a major economic and health impact throughout the world [1,23,199]. This pathogen is a member of the *Mycobacterium tuberculosis* complex which includes many species and subspecies that cause similar pathologies in a variety of mammalian hosts. The most notable member of the complex is *Mycobacterium tuberculosis*, the most important bacterial pathogen of humans. In contrast to *M. tuberculosis*, which is largely host restricted to humans, *M. bovis* is primarily maintained in bovids, in particular domesticated cattle, although this pathogen can frequently be recovered from other mammals including man [1]. Bovine TB is found in cattle throughout the world and has been reported on every continent where cattle are farmed [200].

Bovine TB has been reduced or eliminated from domestic cattle in many developed countries by the application of a test-and-cull policy that removed infected cattle [1,5,23,32,199-201]. However, in Africa, although bovine TB is known to be common in both cattle and wildlife, control policies have not been enforced in many countries due to cost implications, lack of capacity and infrastructure limitations [5,20,32,201]. In 1998, Cosivi et al. reported of bovine TB "Of all nations in Africa, only seven apply disease control measures as part of a test-and slaughter policy and consider bovine TB a notifiable disease; the remaining 48 control the disease inadequately or not at all" [32]. In the intervening years the situation is not thought to have improved [5]; however, preliminary surveys of bovine TB have been carried out in some African countries [47,52,61,66,69,78,86,88].

The most common epidemiological molecular typing method applied to strains of *M. bovis* is spoligotyping. This method identifies polymorphism in the presence of spacer units in the direct repeat (DR) region in strains of the *M. tuberculosis* complex [53,202]. The DR is composed of multiple, virtually identical, 36-bp regions interspersed with unique DNA spacer sequences of a similar size (direct variant repeat or DVR units). Spacer sequences are unique to the DR region and copies are not located elsewhere in the chromosome [182]. The DR region may contain over 60 DVR units; however, 43 of the spacer units were selected from the spacer sequences of the *M. tuberculosis* reference strain H37Rv and *M. bovis* BCG strain P3, and are used in the standard application of spoligotyping to strains of the *M. tuberculosis* complex [202,203]. The DR region is polymorphic because of the loss (deletion) of single or multiple spacers, and each spoligotype pattern from strains of *M. bovis* is given an identifier by www.Mbovis.org.

Several studies of the DR region in closely related strains of *M. tuberculosis* have concluded that the evolutionary trend of this region is primarily by loss of single DVRs or multiple contiguous DVRs [182,203,204]; duplication of DVR units or point mutations in spacer sequences were found to be rare. The loss of discreet units observed by Groenen et al., [203] led them to suggest that the mechanism for spacer loss was homologous recombination between repeat units. However, a study by Warren et al., [205] suggested that for strains of *M. tuberculosis* insertion of IS6110 sequences into the DR region and recombination between adjacent IS6110 elements were more important mechanisms for the loss of spacer units.

The population structure of the *M. tuberculosis* group of organisms is apparently highly clonal, without any transfer and recombination of chromosomal sequences between strains [1,154,206,207]. In a strictly clonal population the loss by deletion of unique chromosomal DNA cannot be replaced by recombination from another strain and the deleted region will act as a molecular marker for the strain and all its descendants. Deletions of specific chromosomal regions (Regions of Difference or Large Sequence Polymorphisms) have been very successful at identifying phylogenetic relationships in the *M. tuberculosis* complex [1,14,137,144,145,149,150,208,209]. However, because the loss of spoligotype spacer sequences is so frequent, identical spoligotype patterns can occur independently in unrelated lineages (homoplasy) and therefore the deletion of spoligotype spacers may be an unreliable indicator of phylogenetic relationship [1,205].

In samples of *M. bovis* strains from Cameroon, Nigeria, Chad and Mali spoligotyping was used to show that many of the strains had a similar spoligotype pattern which lacked spacer 30, and it has been suggested that strains from these four countries are phylogenetically related [43,47,52,61]. We have extended the previous observations of spoligotype similarities between strains from these countries and confirmed the existence of a unique clonal complex of *M. bovis* all descended from a single strain in which a specific deletion of chromosomal DNA occurred. We have named this clonal complex of *M. bovis* strains African 1 (Af1) and show that this clonal complex is dominant in these four West-Central African countries but rare in east and southern Africa. Extended genotyping, using VNTR, of strains with the most common spoligotype patterns suggests that each of these four West-Central African countries has a unique population structure. Evolutionary scenarios that may have led to the present day distribution of the Af1 clonal complex are discussed.

Results

Strains with spacer 30 absent

It has previously been shown that many strains of *M. bovis* isolated from cattle in Mali, Chad and Cameroon have a spoligotype pattern lacking spacer 30 [43,47,61]. In a small sample of 15 strains from the Ibadan slaughterhouse in south-western Nigeria the spoligotype patterns also lacked spacer 30 [52]. To supplement this observation we spoligotyped another 163 strains from the same Nigerian source (table 5). Only strains isolated from cattle were used in this analysis and throughout this manuscript. All spoligotype patterns of strains from these four countries show the loss of spacer 30 except for two of the 65 strains from Chad (SB1102) and seven of the 20 strains from Mali (spoligotype patterns SB0134 and SB0991). The three most common spoligotype patterns from each of the four countries, representing over 65% of strains from each of these countries, are shown in table 6. In general, over 96% of the 338 strains sampled from these countries lacked spacer 30 in their spoligotype pattern (table 5 and appendix 3).

Chad			Cameroon			Nigeria			Mali		
Spoligotype	Frequency	%	Spoligotype	Frequency	%	Spoligotype	Frequency	%	Spoligotype	Frequency	%
SB0944	26	40	SB0944	47	62.7	SB0944	82	46.1	SB0300	8	40
SB1025	13	20	SB1461	8	10.7	SB1027	25	14	SB0134*	6	30
SB0951	7	10.8	SB1460	6	8	SB1025	8	4.5	SB1410	2	10
SB1027	4	6.2	SB0951	5	6.7	SB1421	6	3.4	SB0944	1	5
SB0952	3	4.6	SB1459	2	2.7	SB0328	5	2.8	SB1411	1	5
SB1098	2	3.1	SB0952	2	2.7	SB1432	4	2.2	SB1412	1	5
SB1102*	2	3.1	SB0955	2	2.7	SB1444	4	2.2	SB0991*	1	5
SB1103*	2	3.1	SB1419	1	1.3	SB0951	3	1.7			
SB1101	2	3.1	SB1462	1	1.3	SB0952	3	1.7			
SB1099	1	1.5	SB1463	1	1.3	SB1420	3	1.7			
SB1100	1	1.5				SB1439	3	1.7			
SB0328	1	1.5				SB1440	3	1.7			
SB1418	1	1.5				SB1099	2	1.1			
						SB1424	2	1.1			
						SB1428	2	1.1			
						SB1430	2	1.1			
						SB1435	2	1.1			
						SB1438	2	1.1			
						SB1445	2	1.1			
						SB1026	1	0.6			
						SB1422	1	0.6			
						SB1423	1	0.6			
						SB1425	1	0.6			
						SB1426	1	0.6			
						SB1427	1	0.6			
						SB1429	1	0.6			
						SB1431	1	0.6			
						SB1433	1	0.6			
						SB1434	1	0.6			
						SB1436	1	0.6			
						SB1437	1	0.6			
						SB1441	1	0.6			
						SB1442	1	0.6			
						SB1443	1	0.6			
Total	65			75			178			20	

* Not a member of the Af1 clonal complex. International names for these spoligotype patterns were assigned by www.Mbovis.org

Table 5. The frequency of spoligotype patterns in the four West-Central African countries.

Country	Spoligotype	1	2	3	4	5	6	7	8	9	10	11	12	13	14	15	16	17	18	19	20	21	22	23	24	25	26	27	28	29	30	31	32	33	34	35	36	37	38	39	40	41	42	43	Frequency (%)
Chad	SB0944	■	■		■	■	■	■	■		■	■	■	■	■	■		■	■	■	■	■	■	■	■	■	■	■	■	■		■	■	■	■	■	■	■	■						40
	SB1025	■	■		■	■	■	■	■		■	■	■	■	■	■		■	■	■	■	■	■	■	■	■	■	■	■	■				■	■	■	■	■	■						20
	SB0951	■	■		■	■	■	■	■		■	■	■	■	■	■		■	■	■	■	■	■	■	■	■	■	■	■	■		■	■	■	■	■	■		■						11
Cameroon	SB0944	■	■		■	■	■	■	■		■	■	■	■	■	■		■	■	■	■	■	■	■	■	■	■	■	■	■		■	■	■	■	■	■	■	■						63
	SB1461	■	■		■	■	■	■	■									■	■	■	■	■	■	■	■	■	■	■	■	■		■	■	■	■	■	■	■	■						11
	SB1460	■	■		■	■	■	■	■		■	■	■							■	■	■	■	■	■	■	■	■	■	■		■	■	■	■	■	■	■	■						8
Nigeria	SB0944	■	■		■	■	■	■	■		■	■	■	■	■	■		■	■	■	■	■	■	■	■	■	■	■	■	■		■	■	■	■	■	■	■	■						46
	SB1027	■	■		■	■	■	■	■		■	■	■	■	■	■		■	■	■	■	■	■	■	■	■	■			■		■	■	■	■	■	■	■	■						14
	SB1025	■	■		■	■	■	■	■		■	■	■	■	■	■		■	■	■	■	■	■	■	■	■	■	■	■	■				■	■	■	■	■	■						4
Mali	SB0300	■	■		■	■		■	■		■	■	■	■	■	■		■	■	■	■	■	■	■	■	■	■	■	■	■		■	■	■	■	■	■	■	■						40
	SB0134	■	■				■	■	■		■	■	■	■	■	■		■	■	■	■	■	■	■	■	■	■	■	■	■	■	■	■	■	■	■	■	■	■						30
	SB1410	■			■	■		■	■		■	■	■	■	■	■		■	■	■	■	■	■	■	■	■	■	■	■	■		■	■	■	■	■	■	■	■						10

* Not a member of the Af1 clonal complex. International names for these spoligotype patterns were assigned by www.Mbovis.org

Table 6. The three most common spoligotype patterns in each of four West-Central African countries.

Identification of a specific deletion - RDAf1

The absence of specific spoligotype spacers can be characteristic of a clonal complex of *M. bovis*; a closely related group of strains all descended from a single common ancestor [1]. However, the loss of spoligotype spacers can independently occur in unrelated lineages (homoplasy) and may misidentify members of a clonal complex [1,205]. Deletions of chromosomal DNA (RDs, or LSPs) are less likely to generate homoplasies provided that the deleted region is not flanked by repetitive sequences that generate identical deletions at high frequency [145].

To identify a suitable phylogenetically informative deletion, chromosomal DNA from two strains from Chad lacking spacer 30 in their spoligotype pattern were applied to an *M. tuberculosis*/*M. bovis* composite amplicon-array and subjected to microarray analysis. As expected, the results of this analysis were compatible with the deletion of RD4, 7, 8, 9, 10, 12 and 13 [145]; however, a previously unreported deletion of approximately 5.3kb was identified in both strains and named RDAf1. In contrast, a strain from Chad with spoligotype spacer 30 present (SB1102) was intact for this region by microarray analysis. Sequencing of the RDAf1 region showed that 5322bp were deleted compared with the chromosomal sequences of *M. bovis* BCG [210] and *M. bovis* AF2122 [211]. The RDAf1 deletion removed Mb0587c - Mb0589c and parts of Mb0586c and Mb0590c (corresponding to Rv0572c-Rv0574c and parts of Rv0571c and Rv0575c in *M. tuberculosis* H37Rv). The flanking regions of the deletion showed no similarity to insertion sequences or repetitive DNA and were not GC rich suggesting that this region was not prone to generating homoplasies. We concluded that RDAf1 could be a suitable marker for a clonal complex of strains present at high frequency in Mali, Cameroon, Nigeria and Chad.

Distribution of RDAf1 in West-Central Africa strains

To determine if the RDAf1 deletion could be used as a marker for an important clonal complex we used a PCR method to survey the state of RDAf1 in all available strains from Chad and Mali and a representative sample of strains from Nigeria and Cameroon, a total of 239 strains (appendix 3).

From Nigeria, 148 strains, including the 15 strains published by Cadmus et al., 2006 [52], were tested as well as a sample of 17 strains representing the population of *M. bovis* from Cameroon. All 165 strains from Nigeria and Cameroon were deleted for the Af1 region. In the sample of 20 strains from Mali, seven strains were intact at the RDAf1 region and these strains had spoligotype patterns SB0134 and SB0991; spacer 30 is present in both these spoligotype patterns. Among the 65 strains from Chad four strains were intact at the RDAf1 region in contrast to the remaining 61 strains from that country; two strains with spacer 30 present (spoligotype pattern SB1102) but also two strains with both spacer 30 and 31 absent (spoligotype pattern SB1103) (table 5). In general, for 250 strains surveyed from these four West-Central African countries, 239 strains 15 were deleted for the RDAf1 region; all 239 strains lacked spacer 30 in their spoligotype pattern (appendix 3).

To confirm that the RDAf1 deletion was identical by descent we sequenced across the RDAf1 deletion boundaries in a total of twenty strains originating from different countries and showing distinct spoligotype patterns of the Af1 complex (appendix 3). In all twenty strains the deletion boundaries of RDAf1 were identical suggesting that this deletion is identical by descent in strains from these four countries.

We concluded that a single clonal complex of *M. bovis* strains, defined by the deletion of RDAf1, and marked by the loss of spacer 30, was present at high frequency in Mali, Chad, Cameroon and Nigeria. We named this *M. bovis* clonal complex African 1 or the Af1 clonal complex.

Af1 in other African countries.

Previously published population surveys of bovine TB strains from Uganda, South Africa and Madagascar have shown that, in general, spacer 30 was present in the spoligotype patterns of strains from these countries suggesting that the Af1 clonal complex was not present at high frequency [69,86,88,110]. To confirm this observation a sample of strains chosen to represent the spoligotype diversity of the population in each country were surveyed by PCR for the status of the RDAf1 region. For representative strains from Burundi (N = 10), Madagascar (N = 8), South Africa (N = 11), and Uganda (N = 13) the RDAf1 region was intact. In most of these strains spacer 30 was present (appendix 3). We also surveyed the status of the RDAf1 region in previously unpublished collections of strains from Algeria (N = 23), Ethiopia (N = 15),

Mozambique (N = 20) and Tanzania (N =14). In most strains spacer 30 was present in the spoligotype pattern and in all strains surveyed the RDAf1 region was intact (appendix 3).

We concluded that the distribution and frequency of strains of the Af1 clonal complex was not uniform throughout Africa; Af1 strains were at high frequency in the four West-Central African countries but were rare or absent in Algeria, Burundi, Ethiopia, Madagascar, Mozambique, South Africa, Tanzania, and Uganda (figure 6).

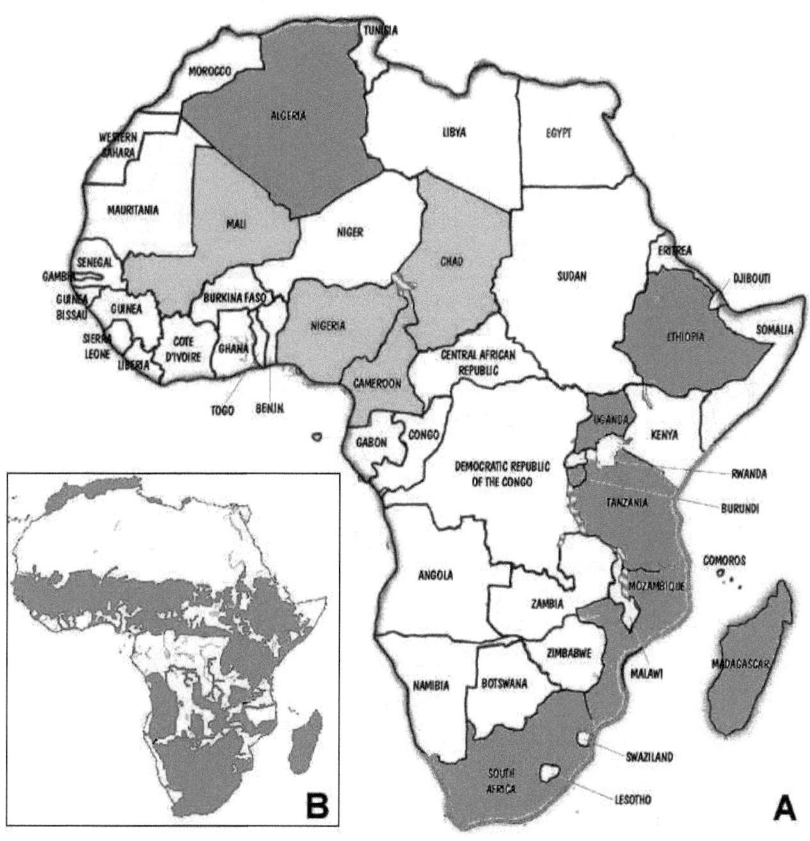

Figure 6. The localization of the African 1 clonal complex of *M. bovis* to West-Central Africa. Panel A: The four West-Central African countries where Af1

strains were found to be dominant are shown in yellow. Countries where Af1 strains were not found are shown in grey. Panel B: Cattle distribution on the continent of Africa (grey shaded area). Modified from Hanotte et al., 2002 [212] with the author's permission.

Geographical localization of genotypes to each country

The population structures of the Af1 strains in Cameroon, Nigeria and Chad are superficially similar; in each of these countries over 90% of the strains included in this study are members of the Af1 clonal complex and strains with spoligotype pattern SB0944 are most common. However, strains from Mali show a distinct difference in population structure compared to the other three West-Central African countries (table 5) although some caution is needed for this interpretation because of the small number of isolates. In Mali, the majority of Af1 strains lack spacer 6 in the spoligotype patterns in addition to the loss of spacer 30; strains of the Af1 clonal complex lacking spacer 6 are rare or absent from Chad, Nigeria or Cameroon. Furthermore, in Mali, members of the Af1 clonal complex make up only 60% of the sampled population and a second group of strains with RDAf1 intact, spacer 30 present but lacking spacers 4 and 5 are also common (spoligotype patterns SB0134 and SB0991, 40% of strains, table 6) [47]. This contrasts with the three other West-Central African countries where strains of the Af1 clonal complex are apparently ubiquitous (table 5). These observations suggest that the population of *M. bovis* in Mali is markedly different from the three other West-Central African countries analyzed here.

To further extend this observation we VNTR typed (ETR-A to F) all available strains with spoligotype pattern SB0944 from Cameroon, Nigeria and Chad to give a genotype for these strains consisting of a combination of spoligotype pattern and 6-loci VNTR pattern. Table 7 shows the genotypes of strains with spoligotype pattern SB0944 from Cameroon, Chad and Nigeria.

Chad		Nigeria		Cameroon	
Genotype*	Frequency	Genotype	Frequency	Genotype	Frequency
SB0944 ; 3 6 5 4* 3 3.1	2	SB0944 ; 5 5 3 4* 3 3.1	22	SB0944 ; 5 5 5 4* 3 2.1	3
SB0944 ; 4 5 5 2 3 3.1	2	SB0944 ; 5 5 5 4* 3 3.1	19	SB0944 ; 5 5 3 4* 3 3.1	2
SB0944 ; 4 5 6 5* 3 3.1	2	SB0944 ; 4 5 5 4* 3 3.1	2	SB0944 ; 5 3 4 4* 4 2.1	1
SB0944 ; 4 7* 5 4* 3 3.1	2	SB0944 ; 5 4 5 4* 3 1.3	2		
SB0944 ; 4 8 5 4* 3 3.1	2	SB0944 ; 5 4 6 4* 3 3.1	2		
SB0944 ; 3 5 3 4* 3 3.1	1	SB0944 ; 5 5 3 4* 3 1.3	2		
SB0944 ; 3 5 5 4* 3 3.1	1	SB0944 ; 3 5 5 4* 3 3.1	1		
SB0944 ; 4 3 5 4* 3 3.1	1	SB0944 ; 4 4 3 4* 3 3.1	1		
SB0944 ; 4 5 6 4* 2 3.1	1	SB0944 ; 5 3 5 4* 3 3.1	1		
SB0944 ; 4 6 5 4* 3 3.1	1	SB0944 ; 5 4 5 4* 2 1.3	1		
SB0944 ; 4 7 5 4* 3 3.1	1	SB0944 ; 5 4 5 4* 3 3.1	1		
SB0944 ; 4 7 6 4* 3 3.1	1	SB0944 ; 5 5 4 4* 3 3.1	1		
SB0944 ; 4 8 5 3* 3 3.1	1	SB0944 ; 5 5 6 4* 3 3.1	1		
SB0944 ; 5 5 3 4* 3 3.1	1	SB0944 ; 5 6 5 4* 3 3.1	1		
		SB0944 ; 6 4 5 4* 3 3.1	1		
		SB0944 ; 6 4 7 4* 3 3.1	1		
Total	19		59		6

*International name for the spoligotype pattern followed by a semi-colon followed by the allele call for the ETR A to F loci. All shared genotypes have been marked.

Table 7. The genotypes of strains with spoligotype pattern SB0944 from Chad, Nigeria and Cameroon.

It is clear from table 7 that strains with the most common spoligotype pattern in Nigeria and Chad (SB0944) differ significantly in VNTR pattern. Only two of the 16 genotypes found in Nigeria are also found in Chad. For Nigerian strains a '5' allele is common at the ETR-A locus (53 of 59 strains) whereas in Chad only one of 19 strains had this allele. Furthermore, the most common genotype in Nigeria (SB0944; 5 5 3 4* 3 3.1), found in 22 of the 59 Nigerian strains, is rare in Chad. The small sample of strains from Cameroon with spoligotype pattern SB0944 also suggest that this population is unique compared to Nigeria and Chad. Four of the six strains genotyped from Cameroon have a 2.1 allele at the ETR-F locus, which is not found in the 78 strains with spoligotype pattern SB0944 from the other two countries from which we were able to generate a full genotype.

The genotypes of strains with the second most common spoligotype pattern in Chad (SB1025) also suggest a difference in the population structure between Nigeria and Chad. Seven of eight strains from Nigeria with spoligotype pattern SB1025 had 5 alleles at both the ETR-A and ETR-B locus whereas eight of ten strains from Chad had 4 and 3 alleles at the ETR-A and B loci respectively (appendix 3).

These observations suggest that the strains sampled from Mali, Chad and Nigeria each have a country specific population structure. That is, given a strain blind from one of these countries it would be possible, with reasonable accuracy, to identify the country of origin from the genotype. The limited data from Cameroon also suggests a country-specific population.

Discussion

We have identified an epidemiologically important clonal complex of *M. bovis* dominant in Mali, Cameroon, Nigeria and Chad and have named this clonal complex African 1 (Af1). This clonal complex is epidemiologically important because it is at high frequency in these four countries; we do not yet know how phylogenetically distinct this clonal complex is from other strains of *M. bovis*. Members of this clonal complex are defined by a 5.3 kb deletion of chromosomal DNA which we have named Region of Difference Af1 (RDAf1). Sequencing of the RDAf1 region in many strains has shown that the deletion boundaries are identical and, in the absence of repetitive elements flanking

RDAf1 and the apparent strict clonality of *M. bovis*, we conclude that this deletion is identical by descent in strains from these four countries. That is, RDAf1 was deleted from the most recent common ancestor of this clonal complex and this region is therefore deleted in all descendents. A definition and summary of the Af1 clonal complex is shown in table 8.

Strains of the Af1 clonal complex can be identified by the loss of spacer 30 in the spoligotype pattern although this characteristic is not necessarily specific. It is theoretically possible for strains with the RDAf1 deletion to have spacer 30 present although we have not yet identified such an isolate. Furthermore, because the loss of spacers in spoligotype patterns can be homoplastic [1,205], strains that are not members of the Af1 clonal complex (RDAf1 region intact) can also lack spacer 30; for example the strains with spoligotype pattern SB1103 from Chad.

African 1 (Af1) clonal complex of M. bovis	
Definition	Presence of deletion RDAf1 (5.3kb, between Mb0586c and Mb0590c)
Spoligotype marker	Absence of spacer 30
Spoligotype signature*	1101111101111110111111111111101111111100000 (SB0944)
Distribution	At high frequency in sub-Saharan West-Central Africa (Mali, Cameroon, Chad and Nigeria)

* The spoligotype signature represents the assumed spoligotype pattern in the progenitor strain of this clonal complex and is shown as a series of 1s and 0s with 1 representing hybridisation to the spacer and 0 representing absence of hybridisation. International name for this spoligotype pattern was assigned by www.Mbovis.org

Table 8. The definition of the Af1 clonal complex.

We have surveyed for the presence of strains of the Af1 clonal complex in small samples of strains from other African countries and have shown that RDAf1-deleted strains are not represented in our samples from Algeria, Burundi, Ethiopia, Madagascar, Mozambique, South Africa, Tanzania, and Uganda (figure 6). Although the number of strains sampled in each of these countries is small, the high frequency of Af1 in the four West-Central African countries supports our conclusion that the Af1 clonal complex is not uniformly distributed throughout Africa but is dominant in sub-Saharan West-Central Africa. Previously published large-scale spoligotype population surveys of *M. bovis* strains from Europe (France, Spain, UK, Italy, Germany, Belgium, Czech

Republic and Portugal) [1,97,98,131,132,146,155,179,187,213-215], the Middle East (Iran) [216], and South and Central America [196,217,218] do not show a high frequency of strains with spacer 30 missing, suggesting that if Af1 strains are present they are not at the high frequency seen in West-Central Africa and supporting our suggestion of geographical localization of the Af1 clonal complex to this region of Africa.

National localization of Af1 genotypes

The population structure of *M. bovis* in Mali is clearly different from the population structure in the three other West-Central African countries (table 5). We extended this observation to show that the genotypes (spoligotype plus VNTR type) of the most common types found in Chad and Nigeria also differ in both frequency and type. However, it could be suggested that these samples are non-representative of the population structure of *M. bovis* in Chad and Nigeria; that is, for the short period when the strains were collected a subset of the total *M. bovis* population present in each country was sampled. However, the strains from Chad were collected over a three year period and the two sets of Nigerian strains, although collected at the same abattoir, were sampled over a year apart; In both these samples from Nigeria the '5' allele at the ETR-A locus was common in contrast to strains from Chad. A second small group of five strains with spoligotype pattern SB0944, collected at an abattoir in southern Chad at least 400km from the collection site of the original Chadian strains in N'Djaména and three years later, was also distinct when compared to Nigerian strains (unpublished data). Taken together these data strongly suggest a distinct, country-specific population structure for *M. bovis* in each of these West-Central African countries. This is notable because it is known that there is extensive movement of cattle, both commercial and transhumance, between Chad, Nigeria and Cameroon, however, our results suggest there is not enough local transmission of extra-national strains of *M. bovis* to obscure the country specific population structure. It is common for livestock sold from Chad to either Cameroon or Nigeria to be directed to large city abattoirs and not resold to mix with the local populations. In a similar way, large cattle populations are sold in Mali for slaughtering in Abidjan (Côte d'Ivoire) and do not mix with local cattle population [219]. The observation of a country specific population structure is somewhat surprising as these West-Central African countries have a century-old, long-distance transhumant livestock production systems and trade routes of cattle between the Sahelian countries and the large coastal cities, which were considered as important determinants of the regional spread of bovine TB [220-222]. More important, our results show that it would be worthwhile for each of these countries to establish simple genotype surveillance of *M. bovis* using spoligotyping and VNTR typing to monitor the import of extra-national strains in a bovine TB control or eradication campaign.

Although our data do not address within-country geographic localization of *M. bovis* genotypes, as is commonly seen in Great Britain [1], we are aware that the previous population survey of *M. bovis* strains in Cameroon suggests a difference in the genotype of strains isolated in the north and the south of the country [61]. This observation implies that it may be worthwhile to carry out more extensive studies of the geographical localization of *M. bovis* genotypes, which, together with an in-depth knowledge of contemporary livestock trade and transhumance routes within Chad, Mali and Nigeria, could lead to the development of useful epidemiological tools (genotype-location maps) to identify key determinants of transmission and to aid the control of bovine tuberculosis in these countries.

The evolution of the Af1 clonal complex

The simplest explanation for the observed distribution and population structure of the Af1 clonal complex throughout these sub-Saharan West-Central African countries is that a single strain or clonal complex of *M. bovis* spread between these four countries in cattle naïve to bovine tuberculosis. The progenitor strain would have had spacer 30 missing and carried the RDAf1 deletion and this would account for the presence of strains descended from this progenitor Af1 strain in each of these four countries (table 8). Country specific population structures could have evolved by drift either during the spread of the Af1 clonal complex between countries (a series of founder events) or subsequently as the population expanded in each country. In Mali, for example, the founding Af1 clone has lost spacer 6 and this subclone has become dominant, while in each of the three other West-Central African countries differences in VNTR profile have evolved. All spoligotype patterns of the Af1 clonal complex in these four

West-Central African countries can be derived from the spoligotype pattern of the progenitor strain by loss of spoligotype spacers. In support of the suggestion of a single introduction of bovine TB and subsequent spread through a population of cattle naïve to bovine tuberculosis it is interesting to note that Alhaji, writing in 1976, reports that bovine tuberculosis first appeared in West Africa in the Cameroons in 1913 and that much of the bovine TB identified in Nigerian abattoirs in the 1940s came from cattle imported from Cameroon [223]. The localization of the Af1 clonal complex to this region of West-Central Africa may have been governed by geographical barriers. To the north of the Af1 range is the Sahara desert and to the South are the forest areas in the Congo. These geographical features reduce both cattle density (figure 6B) and movement and therefore may have limited the spread of Af1 strains. There are no major geographical barriers between Chad, Mali, Cameroon and Nigeria and cattle density throughout the region is fairly uniform (figure 6B). The absence of Af1 in East Africa, particularly in Ethiopia, can be explained by poor trade links between West-Central and East Africa or the prior establishment of an *M. bovis* population in East Africa countries that may have limited the introduction of the Af1 clonal complex. We assume that other countries in this region, especially Niger, will also be dominated by strains of the Af1 clonal complex and that these countries will have country-specific population structures (figure 6). It is noteworthy that a strain of *M. bovis* with spacer 30 missing in the spoligotype pattern has been isolated from a human with pulmonary tuberculosis in Ghana [224].

It may seem unreasonable that a single strain spread between these countries with both a chromosomal deletion and the deletion of a spoligotype spacer. However, we can assume that the population of *M. bovis* that gave rise to the Af1 progenitor was very small (a founder population) and, under those conditions, in a clonal organism, the fixation of otherwise deleterious mutations may be quite common [1]. Whether the progenitor of the Af1 clonal complex evolved in Africa or elsewhere and was subsequently imported to Africa is unknown although this may be resolved when the phylogenetic relationship of the Af1 clonal complex to strains from other countries is determined.

Strains of the Af1 clonal complex have virtually reached fixation in Nigeria, Cameroon and Chad and the suggestion of a single strain spreading throughout the region and subsequently establishing country-specific populations is, to a certain extent, dependent on the absence of *M. bovis* strains in these three countries prior to the spread of Af1. However, if each country had a prior population of *M. bovis* that was replaced by the Af1 clone then it is difficult to explain, without invoking selection, how strains of this clonal complex went to fixation by drift, independently, in three countries. If we assume West-Central African cattle were infected with a prior population of *M. bovis*, then the Af1 clone could have arisen with a selective advantage and spread throughout the region, going to fixation in Nigeria, Cameroon and Chad and replacing the previous population.

It would be tempting to suggest that the RDAf1 deletion generated a phenotype that was selectively advantageous. For some *M. bovis* deletions a phenotype has been described, for example, deletion RD4 induces a truncated form of a phenolic glycolipid [225]; however any selective advantage of this phenotype is unknown. In general, attempts to assign selective advantages to the many other deletions in members of the *M. tuberculosis* complex have not been overly successful. Exceptions are the deletion of the esxAB genes and flanking regions seen in BCG (RD1), *M. microti* (RDMic), and the Dassie bacillus [226-228], and the loss of an immunoregulatory locus from an epidemic strain of *M. tuberculosis* [229]. The loss of the esxAB region in several independent deletion events is strong evidence for an unknown selective advantage for the deletion of this locus. However, before a selective advantage is assumed for any particular chromosomal deletion it should be remembered that in a clonal organism deletions that are not advantageous can easily go to fixation by hitchhiking with a selectively advantageous mutation (SNP or deletion) located elsewhere on the chromosome [1].

Homology searches in the databases at NCBI suggest that the five genes (Mb0586c - Mb0590c) affected by the RDAf1 deletion are conserved among several related *Mycobacterium* species but none of the genes has yet been functionally characterized. However, Mb0589c (Rv0574c) shares homology with the gene for CapA from *Bacillus anthracis* and PgsAA from *Bacillus subtilis*; these proteins are proposed to be part of a complex for poly-gamma-glutamate biosynthesis [230].

In Mali, two major clonal complexes of *M. bovis* have been described [47]. Strains of Af1 make up 60% of the population whereas a second clonal complex, marked by the loss of spacers 4 and 5 and the presence of spacer 30 in the spoligotype pattern, is also present at high frequency (spoligotype SB0134 and related patterns) [47]. This second clonal complex of strains in Mali, provisionally named Af5, has been shown to have spoligotype similarities to strains from Europe [47]. Whether the Af1 clonal complex went to high frequency in Mali by either drift or selection the simplest explanation for the presence of the Af5 clonal complex is that it was introduced after Af1 had established; however we have no evidence to indicate the relative time of establishment of Af1 and Af5 or the phylogenetic relationship between them except that strains of Af5 are not members of the Af1 clonal complex.

Global distribution of clonal complexes of M. bovis

The local dominance of specific clonal complexes of *M. bovis*, identified initially by spoligotype signatures, is becoming a feature of the global population structure of this economically important pathogen [155]. A clonal complex of limited diversity, provisionally called Eu1, dominates in the British Isles and historic trading partners [1] and we also suspect that a second clonal complex of *M. bovis* dominates in East Africa countries (Af2) and a third in Madagascar (Af3).

Identification of these clonal complexes will assist not only the epidemiological analysis and the international movement of *M. bovis* strains in cattle but also in humans [217]. A strain of *M. bovis* isolated in 2005 from a human at the Midlands Regional Centre for Mycobacteriology, Birmingham, UK had the commonest genotype seen in our Nigerian sample (SB0944 ; 5 5 3 4* 3 3.1) [231]. This genotype is quite distinct from the genotypes currently isolated from cattle and humans in the British Isles and we have shown that this strain is deleted for the RDAf1 region and therefore a member of the Af1 clonal complex (unpublished data). It is interesting to note that this patient was born in West-Central Africa. Furthermore, a strain of *M. bovis* with spoligotype pattern SB1025, the second most common spoligotype identified in our sample from Chad, was isolated from a mother and daughter of Canadian origin in France in 2003. These strains were also deleted for the RDAf1 region (unpublished data) and the infection was presumably acquired in Africa. These cases illustrate how the identification and analysis of *M. bovis* clonal complexes and genotypes can be used to suggest possible sources of infection.

Materials and Methods

Bacterial strains

All strains analyzed in this manuscript were isolated from cattle and further information is given in the appendix 3. Of 89 *M. bovis* strains isolated from Chadian cattle, 65 were isolated from animals sampled in the years 2000-2002 at the N'Djaména abattoir [43,142]; 24 additional strains originated from cattle sampled between July and November 2005 at the abattoir of Sarh, in southern Chad approximately 400 km away from N'Djaména. Of 178 *M. bovis* strains isolated from Nigerian cattle, 15 strains isolated at the Bodija abattoir in Ibadan, Nigeria in 2003 were previously published [52] while 163 strains isolated at the same abattoir between April and August 2004 are described in this study for the first time. The 75 strains from Cameroonian cattle were collected in 1989-1990 and 1995-1996 from cattle in different abattoirs and were published previously [61]; a representative subset of 17 strains was used for molecular analyses in this study. The 20 strains from Mali were isolated in March and April 2007 from cattle at the Bamako abattoir and have been previously described [47]. All strains were characterized by spoligotyping, and the majority were subjected to VNTR typing, RD4 typing and RDAf1 typing (appendix 3). Population sample of *M. bovis* from Madagascar (N = 8), South Africa (N = 11), Uganda (N = 13), Burundi (N = 10), Tanzania (N = 14), Ethiopia (N = 15), Mozambique (N = 20) and Algeria (N = 23) were also analyzed by spoligotyping and RDAf1 deletion typing (appendix 3). *M. tuberculosis* H37Rv and *M. bovis* AF2122/97 were included as reference strains in our experiments.

Spoligotyping and VNTR typing

Strains were spoligotyped according to the method of Kamerbeek et al. [53] with minor modifications [52]. VNTR typing targeted the six loci originally described by Frothingham et al. [138] according to the protocol described by Cadmus et al. [52]. Presence of a 24 bp deletion frequently observed in one of the tandem repeats of the ETR-D locus was indicated by a * i.e. 4* (=3 x 77 bp repeats and one 53 bp repeat). The ETR-F locus contains two sorts of tandem repeats of different length (79 bp and 55 bp). We displayed the repeat number of the 79 bp repeats followed by the repeat number of the 55 bp repeats separated by a period. All strains were VNTR typed at VLA, Weybridge, UK.

Microarray analysis

For the microarray analysis, four isolates (No. 86, 111, 486 and 515; appendix 3) were selected from the Chadian *M. bovis* collection, each representing a different spoligotype. The isolates either lacked spacer 30 (strains 111 and 486), spacers 20-22 (strain 515) or a combination of both (strain 86). Approximately 1-4 µg whole genomic DNA for each isolate was extracted as described [232]. The array used in this study contained non-redundant protein coding sequences (CDS) from the two sequenced *M. tuberculosis* strains, H37Rv, CDC1551 and from the sequenced *M. bovis* strain, AF2122/97 (http://bugs.sghms.ac.uk). Random amplification, labeling of genomic DNA and microarray hybridizations were performed as described [233]. GeneSpring 5.0 was used for the data analysis and a cut-off for the normalized test/control ratio of <0.5 was used to create gene deletion lists. Deletions found in regions associated with repetitive elements and insertion sequences, which are known to be prone to deletion events, were disregarded in this study.

Deletion typing

The majority of the strains from Nigeria, Chad, Mali and Cameroon identified as *M. bovis* by spoligotyping were additionally confirmed by the deletion of RD4 [1,145]. The presence or absence of RDAf1 was assessed by multiplex PCR with a set of three primers (RDAf1 primer set A); two primers targeting the flanking regions of RDAf1 (Mb0586c FW: 5'-ACTGGACCGGCAACGACCTGG and Mb0590c Rev: 5'-CGGGTGACCGTGAACTGCGAC) and one primer hybridizing with the internal region of RDAf1 (Mb058xc Int Rev: 5'-CGGATCGCGGTGATCGTCGA). A 350 bp (RDAf1 intact) or a 531 bp PCR product (RDAf1 deleted) was identified by agarose gel electrophoresis. RDAf1 primers and control strain supernatants are available from the corresponding author on request. PCRs contained per reaction 1µl of supernatant of heat-killed mycobacterial cells, a final concentration of 1xHotStartTaq Master Mix (Qiagen), 1µM of primer Mb0586c FW, 0.5 µM of primer Mb0590c Rev and Mb058xc Int Rev and sterile distilled water to a final volume of 20 µl. Thermal cycling was performed with an initial denaturation step of 15 min at 96°C, 30 cycles of 30 sec at 96°C, 30 sec at 65°C and 1 min at 72°C, followed by a final elongation step of 10 min at 72°C. PCR products were separated on a 1% agarose gel.

Sequencing

The RDAf1 deletion junctions of a sample of 20 strains of the Af1 clonal complex and from different countries were sequenced using primers Mb0586c FW and Mb0590c Rev described for RDAf1 deletion typing (appendix 3). The designation of the strains analyzed and the corresponding Gene Bank accession numbers are: 11b: EU887538; 17b: EU887539; 24b: EU887540; 53b: EU887541; 950 gg mam P: EU887542; 806 rein P: EU887543; 57 HPS pm P: EU887544; 526 gg prescap P: EU887545; 208 gg prescap G: EU887547; 18 HPS pm G: EU887548; 81: EU887549; 45: EU887550; 50: EU887551; 52b: EU887552; C1-128: EU887553; C7-3438: EU887554; B15/05: EU887555; 55: EU887556; 86: EU887557; 54: EU887558 (appendix 3).

Acknowledgements

We thank E. J. Vololonirina, M. Okker and K. Gover for excellent technical assistance. This work was funded by the Swiss National Science Foundation (Grant No. 320000-107559), the National Centre of Competence in Research (NCCR) North-South IP-4, Prionics AG (Zurich, Switzerland), ARC, MRC and NRF in South Africa, the Institut Pasteur de Madagascar, by a Swedish International Development Cooperation Agency grant to Mozambique, work in Mali is supported by the Swiss National Centre of Competence in Research (NCCR) North–South: Research Partnerships for Mitigating Syndromes of Global Change, by the Wellcome Trust Livestock for Life and Animal Health in the Developing World initiatives, the University of Ibadan/MacArthur Foundation and by the Department of Environment, Food and Rural Affairs, UK.

Part III

Diagnosis of bovine tuberculosis in Chadian cattle

Comparative assessment of fluorescence polarization and tuberculin skin testing for the diagnosis of bovine tuberculosis in Chadian cattle

Bongo Naré Richard Ngandolo[1]*, Borna Müller[2]*, Colette Diguimbaye-Djaïbe[1], Irene Schiller[3], Beatrice Marg-Haufe[3], Monica Cagiola[4], Michael Jolley[5], Om Surujballi[6], Ayayi Justin Akakpo[7], Bruno Oesch[3], Jakob Zinsstag[2]
* These authors contributed equally to this study

[1]Laboratoire de Recherches Vétérinaires et Zootechniques, N'Djaména, Chad
[2]Swiss Tropical Institute, Basel, Switzerland
[3]Prionics AG, Schlieren-Zurich, Switzerland
[4]Instituto Zooprofilattico dell`Umbria e delle Marche, Perugia, Italy
[5]Diachemix LLC, Grayslake, USA
[6]Canadian Food Inspection Agency, Ontario, Canada
[7]Ecole Inter-Etats des Sciences et de Médecine Vétérinaires, Dakar, Senegal

This article has published in:
Preventive Veterinary Medicine, 2009

Abstract

Effective surveillance of bovine tuberculosis (BTB) in developing countries where reliable data on disease prevalence is scarce or even absent is a precondition for considering potential control options. We conducted a slaughterhouse survey to assess for the first time the burden of BTB in southern Chad. 954 slaughter animals were sampled consecutively and tested using the single intra-dermal comparative cervical tuberculin (SICCT) test, a recently developed fluorescence polarization assay (FPA) and routine abattoir meat inspection after slaughter. Gross visible lesions were detected in 11.3% (CI: 9.4 - 13.5%) of the animals examined and they were mostly located in lymph nodes and the lung. Significantly more Mbororo zebus (15.0%) were affected by lesions than Arab zebus (9.9%; OR = 2.20, CI: 1.41 - 3.41; $p < 0.001$). Altogether, 7.7% (CI: 6.2 - 9.6%) of the animals were tested positive for *Mycobacterium bovis* infection using SICCT according to OIE guidelines. However, receiver operating characteristic (ROC) analysis using *Mycobacterium tuberculosis* complex (MTBC) infected animals as the positive population and lesion negative animals as the negative population, revealed a better SICCT performance if the cut-off value was decreased to > 2 mm. SICCT reactor prevalence rose to 15.5% (CI: 13.3 - 18.0%) and FPA did not perform better than SICCT, when this setting adapted cut-off was applied.

Introduction

Bovine tuberculosis (BTB) is a considerable threat in many respects. It causes economic loss by its effects on animal health and productivity and by international trade restrictions [19]. BTB has also a high impact on animal wellbeing in wildlife populations and hence entire ecosystems [20]. Moreover, infected wildlife serves as an animal reservoir and hampers BTB eradication programs in several countries [234]. BTB is also of concern for public health as it can cause zoonotic disease in humans, e.g. through close contact to the animals or consumption of raw milk [5,32]. In Africa, the disease is present virtually on the whole continent with only very few countries being able to apply control measures due to the lack of financial resources. Moreover, laboratory and technical capacity is very limited in most countries with diagnosis of tuberculosis relying exclusively on microscopy [5,32,45]

In a representative survey, Schelling et al. found 17% of transhumant nomadic cattle to be positive by single intra-dermal comparative cervical tuberculin (SICCT) testing [56]. A subsequent study at the abattoir of N'Djaména in Chad revealed that 7.3% of the animals had gross visible BTB suspect lesions with Mbororo zebu breeds being more affected than Arab zebus. The differential susceptibility was even more significant, when only confirmed *Mycobacterium bovis* infected animals were considered [43]. Spoligotyping [53] demonstrated a homogeneous population structure of the isolated bacteria with the most predominant strains showing the same patterns as previously identified in studies from northern Cameroon and Nigeria [43,52,61].

Current ante mortem diagnosis of BTB mainly relies on SICCT testing, which although imperfect could not yet be replaced by any other more accurate or satisfactory diagnostic method [162]. Also the Interferon-γ (IFN-γ) test (Bovigam®, Prionics) has gained increasing importance for BTB diagnosis in cattle [164].

SICCT and the IFN-γ test are both based on cell mediated immune (CMI) responses against tuberculosis infection. TB in cattle is characterized by an early Th1 type CMI response, whilst humoral immune responses develop as disease progresses. CMI responses can wane and animals become anergic, and SICCT as well as IFN-γ tests have been shown to give false negative results in such disease stages [162,165,235]. Anergic animals are thought to be heavily diseased and highly infective [165]. In low income countries, where control measures are absent, the predicted higher prevalence of such animals might considerably affect disease spread and persistence [162,165,168]. Thus, development of a diagnostic test targeting late stage diseased animals is of specific importance for high-incidence countries with restricted resources for BTB surveillance and control.

Anergic animals may be detected by serological tests if the host's immune response shifts from a predominantly CMI response to an antibody-based response. In this context, a number of tests have been developed, however sensitivity and/or specificity was low compared to SICCT [162]. Fluorescence polarization (FP) constitutes an alternative technique for antibody detection with a shown potential for diagnostic purposes [236]. An assay for the detection of *Mycobacterium bovis* antibodies has been described some years ago, utilizing fluorescein-labeled MPB70 protein as antigen [170,237-239]. The assay has been modified employing a polypeptide-based tracer derived from MPB70 protein, named F-733 [170]. The present study aimed for the first time at estimating the prevalence of bovine tuberculosis of slaughter cattle in southern Chad by the comparative use of ante-mortem and post-mortem methods.

Materials and Methods

Animals

In the absence of a sampling frame, a total of 954 slaughter animals were sampled during three intervals of approximately one month between July and November 2005 at abattoirs in southern Chad. Sample size calculation for diagnostic test comparison was based on a prevalence of 17%, a standard error of the estimate < 10% and a difference in sensitivity to be detected of 10% (sensitivity of the SICCT test was assumed to be 82%, level of confidence of 95%, power of 80%; www.openepi.com, 2004). The far majority (944 animals) was sampled at the abattoir of Sarh and a few (10 animals) at the abattoir of Moundou. The study area was located approximately 500 km from N'Djaména, where a previous slaughterhouse survey was conducted [43]. The animals were raised in a long distance transhumant livestock production system with frequent trans-border movements of herds between the Central African Republic and Chad [222]. Because of the regional farming system, focusing primarily on milk production [180], relatively small amounts of animals (surplus males and old cows) are usually sold from the same herd to different traders, which in turn, sell on their animals to different butchers. Therefore, we assume that the tested animals can be considered a representative sample of animals from a large number of different herds and a big area around southern Chad. However, because of poor documentation and the multiple selling-on the origin of the animals could not be traced. All available animals were subjected to this study. The reticence of some butchers limited the number of animals that could be sampled to approximately one third of the animals slaughtered during our presence at the abattoir. Different zebu breeds were frequently intermixed in the same cattle herds. None of the animals has ever undergone tuberculin skin testing. Four types of phenotypic breeds were encountered: Arab (N = 658), Mbororo (N = 286), Bogolodjé (N = 7) and cross-breeds (N = 3) of which only the former two were used for statistical analysis.

Physical examination of animals

All 954 animals were physically examined before slaughter. Body condition was assessed by assigning one of the following three scores: 1 – good body condition, 2 – bad body condition, 3 – very bad body condition. Classification of the animals was based on the bodyweight and external characteristics (dermatosis, etc.).

Lymph adenopathy was assessed by examination of the left and right pre-scapular and superficial inguinal lymph nodes and assigning different scores of hyperplasia: 1 – normal, 2 – medium hyperplasia of the lymph node, 3 – severe hyperplasia of the lymph node.

Fluorescence polarization assay (FPA)

In our study, a polypeptide-based tracer derived from MPB70 protein, named F-733 was used for FPA in microtiter (GENios Pro) and single tube format (SENTRY 100) [170].

a. GENios Pro/Microtiter plate format:

Blood samples from all 954 animals were subjected to GENios Pro FPA. The assay was performed as direct assay using fluorescein labeled F-733 tracer followed by a confirmatory inhibition assay using fluorescein labeled F-733 tracer in combination with a large excess of unlabeled 733 peptide as described by Jolley et al. [170]. Assay buffer (0.01 M sodium phosphate pH 7.5, containing 9 g/l sodium chloride, 1 g/l sodium azide and 4 g/l lithium dodecyl sulfate) of a volume of 100 µl and 100 µl of serum sample were applied into microtiter plate wells (96-well flat bottomed black microtiter plates, Greiner, catalogue no. 7.655 209), mixed thoroughly for 5 min on a microplate shaker and incubated for 25 min at room temperature. After background reading, 10 µl of F-733 tracer solution (125 nM in 0.01% sodium phosphate buffer, pH 7.5, containing 9 g/l sodium chloride, 1 g/l sodium azide and 100 mg/l bovine gamma globulin) was added, mixed thoroughly for 5 min on a microplate shaker and incubated at room temperature for further 5 min. Then the background-subtraced FP of the tracer was determined. For the confirmatory assay, the assay was performed using unlabeled peptide (10 µl, added to 90 µl of buffer) instead of 100 µl of buffer. The samples were analyzed in duplicates without inhibitor and retested in duplicate with inhibitor. All FP measurements were performed with the GENios Pro™ (Tecan AG) microplate reader.

b. SENTRY 100/Single tube format:

Blood samples from all 954 animals were subjected to SENTRY 100 FPA; however, valid test results were available for 953 animals. Assay buffer (0.01 M sodium phosphate pH 7.5, containing 9 g/l sodium chloride, 1 g/l sodium azide and 1 g/l lithium dodecyl sulfate; 900 µl) and 100 µl of the serum sample were added to borosilicate glass test tubes (Culture tubes, Durex, 10x75 mm; VWR cat. no: 47729-568) and mixed vigorously for 5 sec. After 2 h of equilibration at room temperature, background reading was performed. 10 µl of F-733 tracer solution (125 nM in 0.01% sodium phosphate buffer, pH 7.5, containing 9 g/l sodium chloride, 1 g/l sodium azide and 100 mg/l bovine gamma globulin) was added. The reagents were again mixed thoroughly for 5 sec and equilibrated for 30 min at room temperature. FP measurements were made using the SENTRY 100 fluorescent polarization instrument (Diachemix LLC, USA). Calibration was performed according to manufacturer's instructions. Three negative and one positive control were included in each batch of sera tested.

Comparative intra-dermal tuberculin skin testing

Valid SICCT testing results were available for 930 animals. To be able to perform SICCT on slaughter animals, an arrangement was made with the slaughterhouse management to maintain animals three day prior to slaughter in the animal confinement area of the slaughterhouse. The skin test was performed using bovine (50'000 IU/ml) and avian (25'000 IU/ml) tuberculin produced by the Istituto Zooprofilattico Umbria e Marche, Perugia, Italy. SICCT testing was effected by intra-dermal injection of 0.1 ml of bovine "Purified Protein Derivative" (PPD-B) and 0.1 ml of avian PPD (PPD-A) on 2 different sites of the mid-neck previously shaved, which gives a dose of 5 000 IU of PPD-B and 2 500 IU of PPD-A per animal. Reading of the skin swelling was performed with a caliper 72 h later. The outcomes were interpreted as recommended by the OIE (Council Directive 64/432/EEC [163]): The result was considered positive if the increase in skin thickness at the PPD-B site of injection was more than 4 mm greater (> 4 mm) than the reaction shown at the site of the PPD-A injection, inconclusive if the reaction to PPD-B was between 1-4 mm greater than to PPD-A and negative if the reaction was equal or less for PPD-B.

Meat inspection

After slaughter, all 954 animals underwent standard meat inspection including organ and lymph node palpation, visual inspection and incision of organs and lymph nodes according to standard procedures [240]. Meat inspection was done by local meat inspectors at the abattoirs of Sarh and Moundou, Chad. Gross visible lesions were detected in altogether 108 of the 954 sampled animals.

Specimen collection, processing and culture

From all 108 animals, which exhibited gross visible lesions during standard meat inspection, lesion containing tissue specimens from all visibly affected organs and lymph nodes were collected and transported on ice to the Chadian National Veterinary and Animal Husbandry Laboratory (Laboratoire de Recherches Vétérinairies et Zootéchniques de Farcha) and stored at -20°C prior to processing. Extracted lesions were homogenized as previously described [43]. The samples were decontaminated with N-acetyl-L cysteine sodium hydroxide (0.5% NALC 2% NaOH) and inoculated into two Middlebrook 7H9 medium flasks containing OADC and PANTA (polymyxin, amphotericin B, nadilixic acid, trimethoprim, azlocillin) and either glycerol (0.75%) or pyruvate (0.6%). Samples were put in culture until growth was detected or at least for 8 weeks. Presence of acid-fast bacilli (AFB) was examined by Ziehl-Neelsen staining and microscopy. Manipulations were carried out as described previously [43].

Culture inactivation and DNA extraction

Bacterial growth was detected in cultures of 102 animals; cultures of 50 animals showed presence of AFB by Ziehl-Neelsen staining. An aliquot of the corresponding cultures was subjected to heat-killing. For each culture containing AFB, 0.5 ml of the bacterial suspension was centrifuged for 5 min at maximum speed, the supernatant was removed and the pellet re-suspended in 0.5 ml of sterile distilled water. The samples were inactivated by incubation at 95-100°C for 15 min in a boiling water bath and stored at -20°C until further processing. Samples were shipped to Switzerland for genetic identification of *Mycobacterium tuberculosis* complex (MTBC) and non-tuberculous mycobacteria (NTM) strains. DNA extraction from thermo-lysates occurred by means of the InstaGene™ Matrix (catalogue no. 732-6030; Bio-Rad).

Light Cycler® PCR

DNA extracts from the 50 animals showing growth of AFB in culture were subjected to Light Cycler® PCR for identification of MTBC and NTM strains as previously described by Lachnik et al. [241] with some modifications. A PCR-mix was made containing 2.5 mM $MgCl_2$, a pair of primers at a concentration of 0.5 µM, each and two pairs of Fluorescence Resonance Energy Transfer (FRET) probes at a concentration of 0.2 µM, each, for the detection of mycobacteria and MTBC species, respectively. 0.1 U of Uracil-N-Glycosylase (UNG) was added for the prevention of carry-over contaminations and the mix was incubated at room temperature for 30 min. 13 µl of the PCR-mix were combined with 2 µl of a commercially available ready-to-use hot start reaction mixture (Light Cycler® Fast Start DNA Master Hybridization Probe; catalogue no. 03 003 248 001; Roche Diagnostics) containing Fast Start Taq polymerase, reaction buffer, deoxynucleoside triphosphates and 10 mM $MgCl_2$. 5 µl of DNA extract was mixed with 15 µl of the amplification mix in a glass capillary. The amplification program began with a denaturation step of 10 min at 95°C, followed by 60 cycles of PCR, with 1 cycle consisting of denaturation (10 s at 95°C), “touchdown” annealing (10 s at 60°C), and extension (45 s at 72°C). Samples were thereupon cooled for 2 min at 40°C.

Primers and probes

Following primers and probes were used for DNA amplification and hybridization: A 1,000-bp fragment was amplified using Mbak-f283 (5'-GAG TTT GAT CCT GGC TCA GGA-3') (sense) and Mbak-r264 (5'-TGC ACA CAG GCC ACA AGG GA-3') (antisense) primers. Following FRET probes were used: for the detection of MTBC species, Mbak TB-F (5'-TCC CAC ACC GCT AAA GCG CTT TCC-3' fluorescein) (antisense) as an anchor probe and Mbak TB-705 (5' Light Cycler Red 705-CCA CAA GAC ATG CAT CCC GTG GTC C- 3') (antisense) as a sensor probe, and for the detection of *Mycobacterium* spp., Mbak GE-F (5'-CTT AAC TGT GAG CGT GCG GGC GAT ACG G-3' fluorescein) (sense) as an anchor probe and Mbak GE-640 (5' Light Cycler Red 640-CAG ACT AGA GTA CTG CAG GGG AGA CTG G-3') (sense) as a sensor probe. As indicated, sensor probes were labelled with Light Cycler Red 640 or Light Cycler Red 705 as an acceptor for FRET, and all anchor probes were labelled with fluorescein.

Statistical analysis

Logistic regression was done with Intercooled Stata version 9.2 (StataCorp LP) Statistically significant covariates for lesion occurrence, were identified by multiple logistic regression.

Non-parametric receiver operating characteristic (ROC) analyses were done in Stata. The roctab and roccomp procedures were used for the sensitivity and specificity calculations for each possible cut-off point, for the calculation of the area under the ROC-curve (AUC), for the comparison of AUCs between the tests and for the production of the ROC plots.

Ideal cut-off values for the diagnostic tests were defined as the points from the ROC plot with the largest distance from the diagonal line (sensitivity = 1 - specificity) [242]; this corresponds to the point with the largest Youden index (J = sensitivity + specificity - 1) [243]. For practical reasons, for GENios Pro and SENTRY 100, the next multiple of ten with the same sensitivity was considered as ideal cut-off and used for test evaluation. For cut-off selection using the misclassification-cost term (MCT), the point with the smallest MCT value ($MCT = (C_{FN}/C_{FP})P(1 - Se) + (1 - P)(1 - Sp)$) was chosen, with C_{FN} and C_{FP}

being the cost of false-negative and false positive diagnosis, respectively and P being the disease prevalence in the target population [243]. We were unable to accurately estimate C_{FN}/C_{FP} but the cost of false-negative diagnosis is likely to exceed the cost of false positive diagnosis. Therefore, MCT values for each possible cut-off point and different ratios of C_{FN}/C_{FP} were calculated and compared, assuming a disease prevalence of 10% and 15%.

Results

Animals and lesions

The animal population sampled at abattoirs in southern Chad (N = 954) consisted mainly of young males and old females in a relatively bad body condition. The majority belonged either to the Arab (N = 658) or Mbororo (N = 286) breed. Interestingly, there was a difference in the sex-ratio between the two zebu breeds (N = 944; $X^2 = 24.1$; $p < 0.001$) with a higher proportion of males in Mbororo compared to Arab cattle. 108 of 954 animals (11.3%; CI: 9.4 - 13.5%) screened had gross visible lesions, which were mostly located in the pre-scapular lymph nodes, the mammary lymph nodes and the lungs (table 9).

	Total number	% of animals with lesions
Animals with lesions	**108**	**100%**
Lymphnode lesions	***98***	***91%***
Prescapular lymph nodes	64	59%
Mammary lymph nodes*	37	34%
Head associated	8	7%
Popliteal lymph nodes	1	1%
Organ lesions	***22***	***20%***
Lung	17	16%
Liver	8	7%
Others	3	3%

* Altogether, 29 (36.7%) of the female animals showed visible mammary lymph node lesions

Table 9. Distribution of gross visible lesions.

In a multiple logistic regression model with lesion occurrence as outcome variable, age and breed were identified as statistically significant explanatory variables (table 10). All visible lesions were put in culture; AFB could be detected in cultures of 50 animals. Using real-time PCR, NTM and MTBC strains could be detected in cultures of 13 and 20 animals, respectively; 3 of altogether 30 animals with confirmed mycobacterial infections showed a mixed infection of NTM and MTBC strains.

Explanatory variable	OR	95% CI	p
Sex	0.97	0.57 - 1.64	0.9
Breed	2.20	1.41 - 3.41	<0.001
Age	1.24	1.11 - 1.38	<0.0001
Body condition	1.02	0.66 - 1.59	0.93
Lymph adenopathy	0.66	0.43 - 1.00	0.05

Pseudo R^2 = 0.0467

Table 10. Logistic model for lesion occurrence.

SICCT

Diagnostic tests were evaluated on the basis of confirmed MTBC infections and lesion occurrence. Using lesion positive and lesion negative animals as positive and negative population, respectively, the ability of SICCT to detect gross visible lesions was assessed. Performance of SICCT for the detection of lesions was poor, reflected by a low AUC (0.60; CI: 0.54 - 0.66; table 11). Using an alternative gold standard definition, confirmed MTBC infected animals were used as the positive population and lesion negative animals were used as the negative population. In this case, SICCT achieved better results (AUC = 0.80; CI: 0.71 - 0.88; table 11; figure 8). We defined the point from the ROC curve with the largest distance from the diagonal line (sensitivity = 1-specificity) as ideal cut-off point [242]. Using this approach, we estimated the best cut-off value to be an increase in skin fold thickness greater than 2 mm (> 2 mm). Sensitivity and specificity of SICCT using our suggested cut-off value and the OIE standard cut-off were 65.0% (CI: 43.3 - 81.9%) / 86.7% (CI: 84.2 - 88.9%) and 20.0% (CI: 5.7 - 43.7%) / 93.1% (CI: 91.1 - 94.6%), respectively (table 11). We found a tuberculin reactor prevalence of 7.7% (CI: 6.2 - 9.6%) for the OIE cut-off and 15.5% (CI: 13.3 -18.0%) for our suggested cut-off (> 2 mm). Mbororo zebus showed for both cut-off values significantly more often a positive SICCT outcome than Arab zebus (N = 920; X^2 = 4.65 and 4.03, $p < 0.05$). Average

SICCT reactor prevalence was already high for animals aged two years or less (5.2% or 10.4% depending on the cut-off used; figure 7). SICCT reactor prevalence continuously increased with age and reached its top in animals aged 5 or 6 years (12.6% or 23%). With higher age, prevalence decreased again (figure 7).

FPA

We evaluated two different FPA methods for the diagnosis of BTB, which were designated GENios Pro and SENTRY 100 depending on their distinct detection system. Low AUCs illustrated that both FPA tests were inappropriate for the detection of animals with gross visible lesions (table 11). ROC analysis for the detection of MTBC infected versus lesion negative animals suggested that the best cut-off point was at 43.3 ΔmP for GENios Pro and between 13.2 ΔmP and 26.3 ΔmP for Sentry100 (figure 8). We therefore classified all animals with a GENios Pro result higher or equal to 40 ΔmP (≥ 40 ΔmP) and animals with a SENTRY 100 result of 20 ΔmP or more (≥ 20 ΔmP) as test positive. Using these cut-off values, sensitivity and specificity of GENios Pro and SENTRY 100 were 50.0% (CI: 29.9 - 70.1%) / 89.8% (CI: 87.6 - 91.7%) and 30.0% (CI: 14.5 - 51.9%) / 97.5% (CI: 96.2 - 98.4%), respectively.

	SICCT		GENios Pro	SENTRY 100
Cut-off	> 4mm**	> 2mm	>= 40	>= 20
Apparent prevalence	7.7% (6.2 - 9.6%)	15.5% (13.3 - 18.0%)	11.1% (9.3 - 13.3%)	3.5% (2.4 - 4.8%)
Lesion positive - Lesion negative				
Sensitivity	13.9% (8.6 - 21.7%)	32.4% (24.3 - 41.7%)	18.5% (12.3 - 26.9%)	11.1% (6.5 - 18.4%)
Specificity	93.1% (91.1 - 94.6%)	86.7% (84.2 - 88.9%)	89.8% (87.6 - 91.7%)	97.5% (96.2 - 98.4%)
Positive predictive value	20.8% (13.1 - 31.6%)	24.3% (18.0 - 31.9%)	18.9% (12.6 - 27.4%)	36.4% (22.2 - 53.4%)
Negative predictive value	89.2% (86.9 - 91.1%)	90.7% (88.5 - 92.5%)	89.6% (87.4 - 91.5%)	89.6% (87.4 - 91.4%)
AUC	0.60 (0.54 - 0.66)		0.57 (0.51 - 0.63)	0.58 (0.52 - 0.64)
MTBC positive - Lesion negative				
Sensitivity	20.0% (5.7 - 43.7%)*	65.0% (43.3 - 81.9%)	50.0% (29.9 - 70.1%)	30.0% (14.5 - 51.9%)
Specificity	93.1% (91.1 - 94.6%)	86.7% (84.2 - 88.9%)	89.8% (87.6 - 91.7%)	97.5% (96.2 - 98.4%)
Positive predictive value	6.6% (1.8 - 15.9%)*	10.7% (6.3 - 17.4%)	10.4% (5.8 - 18.1%)	22.2% (10.6 - 40.8%)
Negative predictive value	98.0% (96.7 - 98.7%)	99.0% (98.0 - 99.5%)	98.7% (97.6 - 99.3%)	98.3% (97.2 - 99.0%)
AUC	0.80 (0.71 - 0.88)		0.67 (0.52 - 0.82)	0.70 (0.58 - 0.82)

* 95% binomial exact confidence intervals are indicated if (estimated value) x (sample size) ≤ 5, otherwise Wilson confidence intervals are shown
** OIE standard cut-off

Table 11. Test performance and characteristics for SICCT, GENios Pro and SENTRY 100.

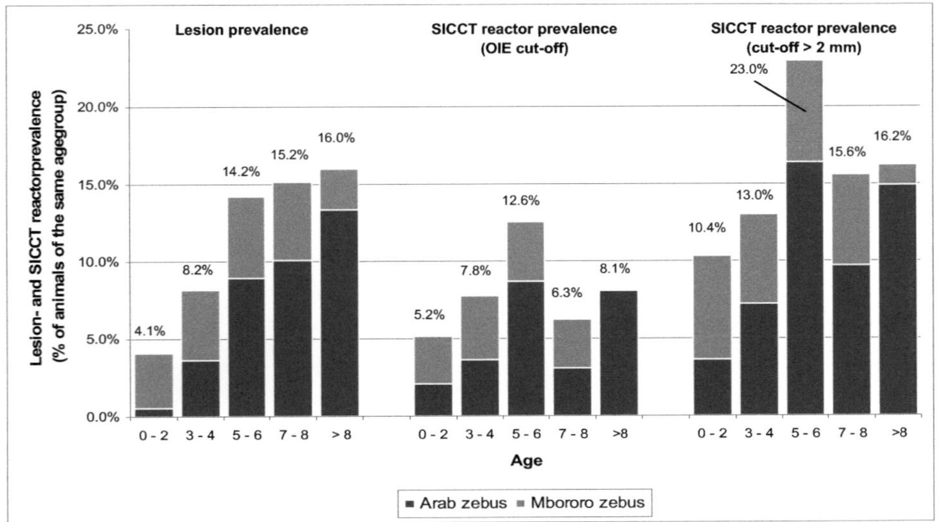

Figure 7. Lesion prevalence and SICCT reactor prevalence for different age groups, using the OIE and cut-off > 2 mm. The prevalence is given in percent of the number of animals of the same age group. The total percentage of positive animals is indicated and the contribution of Arab (dark grey) and Mbororo (light grey) zebus is displayed.

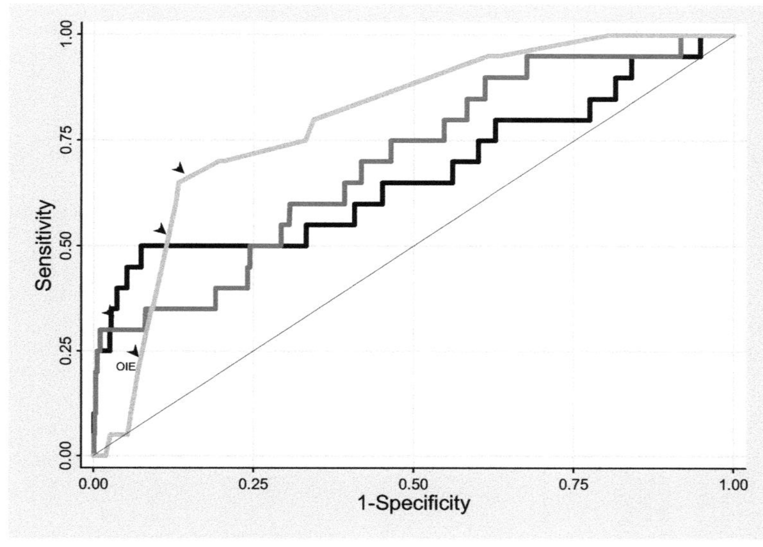

Figure 8. ROC curves for three diagnostic tests: GENios Pro (black), SENTRY 100 (dark grey) and SICCT (light grey). Animals with confirmed MTBC infections were considered positive and animals without lesions negative. Arrowheads correspond to the cut-off points selected in this study (the point corresponding to the OIE standard cut-off for SICCT is indicated).

Comparison of FPA and SICCT

Comparison of SICCT, GENios Pro and SENTRY 100 on the basis of their ROC plots (figure 8) and associated AUCs (table 11) suggested that SICCT generally performed better than both FPA tests. However, the confidence intervals for all AUCs were overlapping and no statistically significant differences could be identified. In fact, most of the confidence intervals for the sensitivities, specificities, positive predictive values (PPV) and negative predictive values (NPV) of the different tests were overlapping and did rarely show any statistically significant differences (table 11). This was most likely due to the small sample size for the confirmed MTBC infected animals (N = 20). Although not at statistically significant levels, our observed data suggested higher sensitivities of the FPA tests compared to SICCT at the OIE cut-off for only somewhat smaller (GENios Pro) or even higher (SENTRY 100) specificities (table 3). SICCT test performance was considerably improved when our suggested cut-off (> 2 mm) was used (table 11, figure 8). However, from the comparison of the test sensitivities for fixed specificities (table 12) and the ROC curve (figure 8) it appeared that FPA performed better than SICCT in the high specificity areas.

		Specificity			
		MTBC infected - Lesion negative			
		80%	85%	90%	95%
Sensitivity	SICCT	70.0%	66.4%	40.3%	5.0%
	Geniospro	50.0%	50.0%	50.0%	40.0%
	Sentry100	40.0%	35.0%	35.0%	30.0%

Predictions derived from actual data

Table 12. Test sensitivities for fixed specificities.

Discussion

Animals and lesions

This is the first ever carried out study on the occurrence of bovine tuberculosis in southern Chad. The region is bordering on northern Cameroon and the

Central African Republic. The animal population encountered at the abattoir is as would be expected for a cattle production system focusing on milk and herd growth, with priority for slaughter given to surplus males and old cows. The different proportion of males in the two zebu breeds could indicate different livestock management strategies for different breeds.

In our previous study in N'Djaména the overall proportion of lesions was significantly lower compared to the present study in Sarh (7.3%, CI: 6.8%-7.8% vs. 11.3%, CI: 9.4% - 13.5%) [43]. Also, the percentage of Mbororo animals encountered was lower in N'Djaména (26.0%; CI: 25.1% - 26.8%) compared to Sarh (30.0%; CI: 27.2% - 33.0%). Since Mbororo animals showed a significantly higher lesion prevalence in both settings, the higher abundance of Mbororo cattle in Sarh may at least in part account for the different overall prevalence between the two settings. Surprisingly and very unlike published in other surveys [63,64,168], routine abattoir meat inspection at Sarh abattoir identified the pre-scapular lymph nodes to be most often affected in animals with lesions (table 9). This suggested that many of these lesions did not result from infection with *M. bovis*. Indeed, SICCT testing provided a sensitivity of only 32.4% (CI: 24.3 - 41.7%; using cut-off > 2 mm) for the detection of animals with lesion but performed significantly better 65.0% (CI: 43.3 - 81.9%) for the detection of MTBC infected animals (table 11). Moreover, in 10 out of 30 mycobacterial infections detected by real-time PCR, NTM were detected without co-infecting MTBC strains. Taken together, these results suggest that a considerable amount of lesions could have been caused by other bacteria than *M. bovis*. Similarly, in two other recent studies from Chad and Uganda, NTM were isolated from more than 40% of the animals exhibiting lesions [58,69]. Considering the unusual frequency of pre-scapular lymph node lesions detected we were in particular interested whether *M. bovis* and other lesion causing bacteria might have infected distinct cattle organs. We re-evaluated the performance of SICCT for the detection of lesions stratified by the lesion location. Although not at statistically significant levels, SICCT performance for the detection of lung lesions was better than its ability to detect pre-scapular lymph node lesions (AUC = 0.72, CI: 0.58-0.68 vs. AUC = 0.57, CI: 0.50-0.65). This suggested that such lesion causing non-MTBC bacteria might more often affect locations outside the lungs. Similar findings were also reported in a recent study from Mali [47].

A relatively high amount of mammary lymph nodes were detected in our survey. This is of public health concern as it could also indicate an increased risk of zoonotic transmission through raw milk consumption. The finding could also explain the relatively high amount of young animals with lesions (figure 7).

SICCT and FPA

Our results suggest that the optimal cut-off for SICCT (> 2 mm) in our setting is lower than the OIE standard cut-off. This finding is of practical relevance and demonstrates the importance of optimizing test cut-offs in the target population. In fact, similar results were found in a study in Ethiopia [65] and are likely to apply to other countries in sub-Saharan Africa. In SICCT reactor prevalence studies in Uganda and Tanzania, lower cut-offs than the OIE standard cut-off have been used, however without detailed justification [33,36]. Using our adapted cut-off, reactor prevalence rises to 15.5% (CI: 13.3 - 18.0%; table 10), indicating a previous underestimation of the disease prevalence in Chad as former studies in other regions of the country were applying higher cut-offs for test interpretation [56,57].

We defined the ideal cut-off as the point from the ROC plot with the largest distance from the diagonal line (sensitivity = 1-specificity) [242,243]. By using this definition, equal weights were given to sensitivity and specificity and it facilitated comparison of different diagnostic tests [243]. In order to take disease prevalence and the cost of misclassifications into account the calculation of the misclassification-cost term (MCT) can be useful for cut-off selection [243]. We were not able to accurately quantify the cost of false negative (C_{FN}) and false positive (C_{FP}) diagnosis as required for this method. However, because a single infected and undetected cow can infect many others and because the economic losses associated with BTB are mostly due to losses in meat production, milk production and increased reproduction efforts [19] we can suppose that the cost of a false negative diagnosis is likely to exceed the cost of a false positive result by several folds. Assuming a disease prevalence of 15% (10%) for the calculation of the MCT, the same ideal cut-off point (> 2 mm) would be identified for SICCT if the ratio of C_{FN}/C_{FP} was between 1.2 and 7 (2 and 11). The selected cut-off values for both FPA tests also gave satisfactory results for these estimates. This suggests that our chosen cut-off values may be acceptable for a broad range of reasonable C_{FN}/C_{FP} ratios. In-depth cost analysis is required to address potential benefits from reduced cut-off values in Africa.

It could be argued that the lesion negative animal population used for diagnostic test evaluation may still contain *M. bovis* infected cattle since no tuberculous lesions are visible at early stages of BTB. Consequently, such animals could have distorted our analysis. In order to increase the certainty that the lesion negative animal population indeed was uninfected we repeated all ROC analyses with a negative animal population that did not show lesions and that in addition showed a ΔPPD value (ΔPPD-B - ΔPPD-A) equal to or less than zero (≤ 0) (for repeated ROC analysis of FPA) or that was in addition negative for both serological tests (for repeated ROC analysis of SICCT). In doing so, 303 and 96 animals were additionally removed from the lesion negative population, respectively. For all three diagnostic tests, the shape of the ROC curve did not distinctly change compared to the initial analysis and AUCs differed by less than 1% (data not shown). This suggests that possible *M. bovis* infected lesion negative animals were present at a negligible frequency and did not falsify our analyses.

The low number of PCR confirmed infections with MTBC strains (N = 20) was problematical for the evaluation of the diagnostic tests. Consequently, AUCs and other performance measures of the different tests differed rarely at statistically significant levels. This is a frequent problem in diagnostic test evaluation. To overcome the problem of the unknown true disease status of lesion negative animals and the low number of confirmed MTBC infected animals, other statistical methods such as latent class analysis may offer alternative approaches for diagnostic test evaluation [244]. Latent class analysis was beyond the remit of this study; however, we are currently trying to develop an approach for Bayesian ROC estimation of multiple tests in absence of a gold standard test.

In a previous study of Jolley et al. [170], GENios Pro results were repeated for positive and suspect positive animals using an inhibition assay (also described in the Materials and Methods section). In our analysis, application of this confirmatory inhibition test lead to an increase of specificity to 99.4% for the detection of lesions at the cost of a lower sensitivity of 10.4% for the detection of lesions and 42.1% for the detection of confirmed MTBC infected animals. Jolley et al. have also shown a high specificity of the test close to 100% but higher values for the sensitivity (61.5% for the detection of PCR confirmed *M. bovis* infections). This discrepancy could be due to a relatively small sample size in both studies [170].

We were interested to see whether confirmed MTBC infected but SICCT anergic animals could be detected by FPA. Using the OIE cut-off and our suggested cut-off, 4/20 and 13/20 MTBC infected animals were detected by SICCT leaving a considerable amount of 16/20 or 7/20 SICCT anergic animals, respectively. GENios Pro detected 2 of the 16 or 7 anergic infected animals and SENTRY 100 only 1. Although the number of confirmed cases was too small to allow any conclusions, there was no indication that anergic animals were more likely to be detected by FPA.

We would like to encourage further research and development for BTB diagnosis in Africa as most of the current work mainly aims at improving diagnostic techniques for industrialized countries. Yet, the majority of cattle affected by BTB originate from the developing world, which can not apply control measures [32]. Current research into improved diagnostic tests does therefore probably not efficiently target the major global burden of BTB.

Acknowledgements

We would like to thank Prof. Erik C. Böttger, Dr. Boris Böddinghaus, Dr. Burkhard Springer and the technicians of the Swiss National Centre for Mycobacteria in Zurich for providing technical support and laboratory facilities. We are indebted to the cattle holders who were collaborating with us within this project. Our work has received financial support from the Swiss National Science Foundation (project no. 107559).

Bayesian Receiver Operating Characteristic Estimation of multiple Tests for Diagnosis of Bovine Tuberculosis in Chadian Cattle

Borna Müller[1], Penelope Vounatsou[1], Bongo Naré Richard Ngandolo[2], Colette Diguimbaye-Djaïbe[2], Irene Schiller[3], Beatrice Marg-Haufe[3], Bruno Oesch[3], Esther Schelling[1], Jakob Zinsstag[1]

[1]Swiss Tropical Institute, Basel, Switzerland
[2]Laboratoire de Recherches Vétérinaires et Zootechniques, N'Djaména, Chad
[3]Prionics AG, Schlieren-Zurich, Switzerland

This article has been published in:
PLoS ONE, 2009

Abstract

Background:

Bovine tuberculosis (BTB) today primarily affects developing countries. In Africa, the disease is present essentially on the whole continent; however, little accurate information on its distribution and prevalence is available. Also, attempts to evaluate diagnostic tests for BTB in naturally infected cattle are scarce and mostly complicated by the absence of knowledge of the true disease status of the tested animals. However, diagnostic test evaluation in a given setting is a prerequisite for the implementation of local surveillance schemes and control measures.

Methodology/Principal Findings:

We subjected a slaughterhouse population of 954 Chadian cattle to single intra-dermal comparative cervical tuberculin (SICCT) testing and two recently developed fluorescence polarization assays (FPA). Using a Bayesian modeling approach we computed the receiver operating characteristic (ROC) curve of each diagnostic test, the true disease prevalence in the sampled population and the disease status of all sampled animals in the absence of knowledge of the true disease status of the sampled animals. In our Chadian setting, SICCT performed better if the cut-off for positive test interpretation was lowered from > 4 mm (OIE standard cut-off) to > 2 mm. Using this cut-off, SICCT showed a sensitivity and specificity of 66% and 89%, respectively. Both FPA tests showed sensitivities below 50% but specificities above 90%. The true disease prevalence was estimated at 8%. Altogether, 11% of the sampled animals showed gross visible tuberculous lesions. However, modeling of the BTB disease status of the sampled animals indicated that 72% of the suspected tuberculosis lesions detected during standard meat inspections were due to other pathogens than *Mycobacterium bovis*.

Conclusions/Significance:

Our results have important implications for BTB diagnosis in a high incidence sub-Saharan African setting and demonstrate the practicability of our Bayesian approach for diagnostic test evaluation.

Introduction

Mycobacterium bovis is the causative agent of bovine tuberculosis (BTB) and belongs to the *M. tuberculosis* complex (MTBC) of bacteria [1]. BTB is a major problem in developing countries, which bear the largest part of the world-wide disease burden and where millions of people are affected by neglected zoonotic diseases such as BTB [5,26,29,32]. The disease causes economic loss by its effects on animal health and productivity and by international trade restrictions [19]. It can also affect health of wildlife [20] and infected wildlife populations serve as reservoirs and hamper disease eradication programs in several countries [234]. Moreover, *M. bovis* infections are of public health concern due to the pathogen's zoonotic potential [5,32].

BTB control and surveillance is scarce in sub-Saharan Africa and mostly limited to abattoir meat inspections. However, the performance of meat inspection is rather poor and depends on the disease stage in which infected animals reside, the accuracy of the carcass examination and the presence of other lesion causing pathogens [63,64,162,245,246]. Recent studies have detected a high proportion of non-tuberculous mycobacteria (NTM) in lesions from Chadian, Ugandan, Ethiopian and Sudanese cattle, suggesting that a considerable amount of lesions detected during abattoir meat inspection of African cattle might be due to other bacteria than *M. bovis* [58,62,67,69].

Current ante mortem diagnosis of BTB mainly relies on the single intra-dermal comparative cervical tuberculin (SICCT) test, which, although imperfect, could not yet be replaced by any other more accurate diagnostic method [162]. SICCT is based on the cell mediated immune (CMI) response against tuberculosis infection. TB in cattle is characterized by an early Th1 type CMI response, whilst humoral immune responses develop as disease progresses. At late disease stages, the CMI response can decrease and SICCT anergic animals can show false negative test results [162,165,247]. Moreover, SICCT performance is influenced by animal exposure to NTM strains as their antigens can cross-react with tuberculin [162]. Serological tests detecting humoral immune responses may be more useful to detect late stage diseased animals. Fluorescence polarization assays (FPA) constitute a technique for antibody detection with a shown potential for diagnostic purposes [236]. An assay for the detection of *M. bovis* antibodies has been described recently [59,170,237-239].

Attempts to evaluate diagnostic tests for BTB in naturally infected cattle in Africa are scarce but a prerequisite for the implementation of surveillance schemes and control measures. Gobena et al. have used detailed post mortem examination to define the BTB disease status of Ethiopian cattle for the evaluation of SICCT in this setting [65]. However, due to the generally low sensitivity and specificity of post mortem meat inspection, its use as a gold standard test is not ideal [246]. We have recently assessed three different tests for the diagnosis of BTB (SICCT and two newly developed FPA methods) in Chadian cattle. Our previous evaluation was also based on a gold standard approach using PCR confirmed MTBC infected and lesion negative animals as the positive and negative population, respectively [59]. Drawbacks of this study were the small number of positive animals and the unknown true disease status of the lesion negative cattle.

Choi et al. [248] developed a Bayesian model for the receiver operating characteristic (ROC) estimation of two diagnostic tests in the absence of a gold standard test. In the present study, we have further extended this model and applied it to evaluate the performance of the diagnostic tests previously assessed by the gold standard approach [59]. Our Bayesian model integrated information from three different diagnostic methods and was independent of a gold standard test; moreover, it allowed us to estimate the true BTB prevalence in the sampled population and the true disease status of each tested animal. Using this information, we could in addition calculate the diagnostic errors of four post-mortem tests (meat inspection, microscopic examination of BTB-like lesions, microscopic examination of derived bacterial cultures and PCR on microscopy positive cultures).

Results

Test results

A total number of 954 sequentially selected slaughter animals from southern Chad were subjected to multiple tests for the diagnosis of BTB. Three ante-mortem tests with continuous numerical outcome values (continuous outcome) were used, namely, SICCT and two recently developed FPA tests termed SENTRY 100 and GENios Pro [59]. Also, four post-mortem tests giving either a positive or negative test result (binary outcome) were applied. These tests were the post-mortem meat inspection, direct microscopy, culture and microscopy and PCR (see materials and methods for details on the applied tests).

Before slaughter, blood samples were collected and animals underwent SICCT testing. Altogether, 8% (CI: 6% – 10%) of the animals tested, reacted positively to SICCT when the official OIE cut-off (> 4 mm; [163]) was used (table 13). Serum extracted from the blood samples was subjected to the FPA tests SENTRY 100 and GENios Pro, for which we have determined most appropriate cut-off values within this study (results shown below; table 13). After slaughter, cattle carcasses underwent meat inspection; lesions suggestive of tuberculosis were isolated from 108 animals (lesion prevalence: 11%; CI: 9% - 14%; table 13). In lesions of 51 animals (47% of animals with lesions; CI: 38% - 57%), acid-fast bacilli (AFB) were observed by direct microscopy (table 13). Culture of lesions and subsequent microscopic examination detected AFB in samples from 50 animals (49% of the animals tested; CI: 39% - 59%; table 13). The microscopy results obtained before and after culture agreed by 86%. In AFB containing cultures of 20 animals MTBC strains could be detected by real-time PCR (table 13). In cultures of 13 animals, NTM strains were detected; three of which showed a mixed infection with MTBC strains.

	No. of animals tested	Outcome	Ante-/post mortem	No. of animals tested pos.	% pos.
SICCT (OIE cut-off > 4 mm) *	930	continuous	ante-mortem	72	7.7%
SICCT (cut-off > 2 mm) *	930	continuous	ante-mortem	144	15.5%
SENTRY 100 (cut-off ≥ 15 ΔmP) *	953	continuous	ante-mortem	62	6.5%
GENios Pro (cut-off ≥ 38 ΔmP) *	954	continuous	ante-mortem	119	12.5%
Meat inspection	954	binary	post-mortem	108	11.3%
Direct microscopy	108	binary	post-mortem	51	47.2%
Culture and microscopy	102	binary	post-mortem	50	49.0%
PCR	50	binary	post-mortem	20	40.0%

* SICCT, SENTRY 100 and GENios Pro results without missing data were available for 929 animals

Table 13. Tests applied for the diagnosis of BTB in Chadian cattle. % pos.: Number of animals tested positively divided by the total number of animals subjected to the respective test.

Model selection

Based on the same data, we have previously reported the evaluation of SICCT, SENTRY 100 and GENios Pro using a subset of animals with either PCR confirmed MTBC infections or no visible lesions [59]. Drawbacks of this approach were the small number of positive animals and the uncertainty about the true disease status of lesion negative animals [59]. The latter is due to the fact that no gross lesions may be observed at early stages of BTB. Here, we describe a Bayesian method for the estimation of the true disease prevalence in the sampled population and the means and variance-covariances of SICCT, SENTRY 100 and GENios Pro test outcomes for the diseased and non-diseased animals. In an initial model we have included data from the post-mortem tests with binary outcomes and attempted to directly estimate their sensitivities and specificities. Prior assumptions and model estimates are indicated in table 14 (models 1A and 1B).

Model estimates for tests with binary outcome were highly sensitive to the priors. We therefore decided to consider solely tests with continuous outcome for Bayesian modeling (model 2A and 2B; table 14). Parameter estimations for these tests did not appear to be sensitive to the prior assumptions and were only marginally different in models 1A, 1B, 2A and 2B (table 14).

Diagnostic test performances

Based on the estimates for the means and variance-covariances of SICCT, SENTRY 100 and GENios Pro test results for the diseased and non-diseased animals in model 2A (table 14), ROC curves were calculated for each test (figure 9) and the most appropriate cut-off for positive test interpretation was defined as the point from the ROC curve with the largest distance from the diagonal line (sensitivity = 1- specificity). For SICCT, a cut-off greater than 2 mm (> 2 mm) appeared to be most appropriate for our setting. For SENTRY 100 and GENios Pro the best cut-off values were determined at 15 ΔmP (≥ 15 ΔmP) and 38 ΔmP (≥ 38 ΔmP), respectively. Using these values, the sensitivities and specificities of the tests were calculated (table 15). The prevalence of *M. bovis* infection in the sampled population was estimated at 8% (CI: 6% - 11%).

In addition to the parameters described above, Bayesian modeling allowed us to compute the latent disease status of the sampled animals. We have used this information from model 2A (table 14) to calculate the sensitivities and specificities of the post-mortem tests with binary outcome to detect modeled *M. bovis* infected animals (table 15). It must be noted, that these estimates refer to the diagnostic performance for our sample, whereas the Bayesian model estimates consider the fact that our sample was a sub-population of the general slaughterhouse population.

Surprisingly, 72% of the animals with gross visible tuberculous lesions detected during standard meat inspection showed a negative result for modeled *M. bovis* infection. Thus, our analysis suggested that 72% of the animals exhibiting tuberculosis-like lesions were infected with other pathogens than *M. bovis*.

Risk factors

We performed logistic regression to identify risk factors for modeled *M. bovis* infection. Univariate logistic regression with modeled *M. bovis* infection as outcome variable and age, sex, animal breed and body condition as explanatory variables identified age and a very bad body condition as risk factors for modeled *M. bovis* infection (table 16). However, in the multiple model, only age turned out to be significantly associated with modeled *M. bovis* infection (table 16). Interestingly, only the presence of organ lesions in general and in particular the presence of lung and liver lesions was significantly associated with modeled *M. bovis* infection (table 17). The presence of lymph node lesions was not associated with modeled *M. bovis* infection (table 17).

Test		Model 1A		Model 1B		Model 2A		Model 2B	
		Prior	Estimate	Prior	Estimate	Prior	Estimate	Prior	Estimate
SICCT	$\mu^{d=0}$	N(0,100)	-0.08 (0.08)	N(0,100)	-0.08 (0.08)	N(0,100)	-0.04 (0.08)	N(10,20)	-0.04 (0.08)
	$\tau^{d=0}$	Ga(1,100)	0.25 (0.02)	Ga(1,100)	0.25 (0.02)	Ga(1,100)	0.24 (0.02)	Ga(1,100)	0.24 (0.02)
	$\mu^{d=1}$	N(0,100)	4.38 (0.55)	N(0,100)	4.38 (0.55)	N(0,100)	4.63 (0.59)	N(10,20)	4.95 (0.63)
	$\tau^{d=1}$	Ga(1,100)	0.04 (0.01)	Ga(1,100)	0.04 (0.01)	Ga(1,100)	0.04 (0.01)	Ga(1,100)	0.04 (0.01)
SENTRY 100	$\mu^{d=0}$	N(0,100)	4.00 (0.22)	N(0,100)	3.99 (0.22)	N(0,100)	4.00 (0.22)	N(10,20)	4.01 (0.22)
	$\tau^{d=0}$	Ga(1,100)	0.03 (0.001)	Ga(1,100)	0.03 (0.002)	Ga(1,100)	0.03 (0.002)	Ga(1,100)	0.03 (0.002)
	$\mu^{d=1}$	N(0,100)	10.52 (2.98)	N(0,100)	10.53 (2.99)	N(0,100)	10.88 (3.27)	N(10,20)	8.83 (2.53)
	$\tau^{d=1}$	Ga(1,100)	0.001 (0.0001)	Ga(1,100)	0.001 (0.0001)	Ga(1,100)	0.001 (0.0001)	Ga(1,100)	0.001 (0.0001)
GENios Pro	$\mu^{d=0}$	N(0,100)	19.52 (0.47)	N(0,100)	19.51 (0.46)	N(0,100)	19.6 (0.46)	N(10,20)	19.62 (0.46)
	$\tau^{d=0}$	Ga(1,100)	0.01 (0.0003)	Ga(1,100)	0.01 (0.0003)	Ga(1,100)	0.01 (0.0003)	Ga(1,100)	0.01 (0.0003)
	$\mu^{d=1}$	N(0,100)	34.56 (3.60)	N(0,100)	34.60 (3.60)	N(0,100)	35.06 (3.92)	N(10,20)	26.43 (3.29)
	$\tau^{d=1}$	Ga(1,100)	0.001 (0.0001)	Ga(1,100)	0.001 (0.0001)	Ga(1,100)	0.001 (0.0001)	Ga(1,100)	0.001 (0.0001)
Meat inspection	m^S	0.5	0.51	0.2	0.23	-	-	-	-
	σ^S	0.1	0.10	0.15	0.16	-	-	-	-
	m^C	0.9	0.94	0.9	0.90	-	-	-	-
	σ^C	0.1	0.02	0.1	0.02	-	-	-	-
Direct microscopy	m^S	0.5	0.51	0.8	0.95	-	-	-	-
	σ^S	0.1	0.10	0.15	0.09	-	-	-	-
	m^C	0.95	0.52	0.95	0.58	-	-	-	-
	σ^C	0.1	0.06	0.1	0.06	-	-	-	-
Culture and microscopy	m^S	0.75	0.76	0.85	0.87	-	-	-	-
	σ^S	0.1	0.10	0.15	0.14	-	-	-	-
	m^C	0.95	0.56	0.95	0.58	-	-	-	-
	σ^C	0.1	0.06	0.1	0.06	-	-	-	-
PCR	m^S	0.75	0.87	0.85	0.97	-	-	-	-
	σ^S	0.1	0.06	0.15	0.04	-	-	-	-
True prevalence	m_π	0.1	0.12	0.05	0.11	0.1	0.08	0.1	0.08
	σ_π	0.1	0.01	0.15	0.01	0.05	0.01	0.05	0.01

Table 14. Priors and model estimates for different parameters. μ^d: Mean diagnostic value for non-diseased (d = 0) and diseased (d = 1) animals, respectively. τ^d: precision of the diagnostic values for non-diseased (d = 0) and diseased (d = 1) animals, respectively. m^S/σ^S, m^C/σ^C, m_π/σ_π : Mean and standard deviation of the test sensitivity, specificity and true disease prevalence, respectively. The normal distribution is parametrized in terms of mean and variance. The Gamma distribution is parametrized in a non-conventional way in terms of mean and variance instead of the shape and scale parameters. Model-based estimates correspond to the posterior mean and standard deviation in brackets.

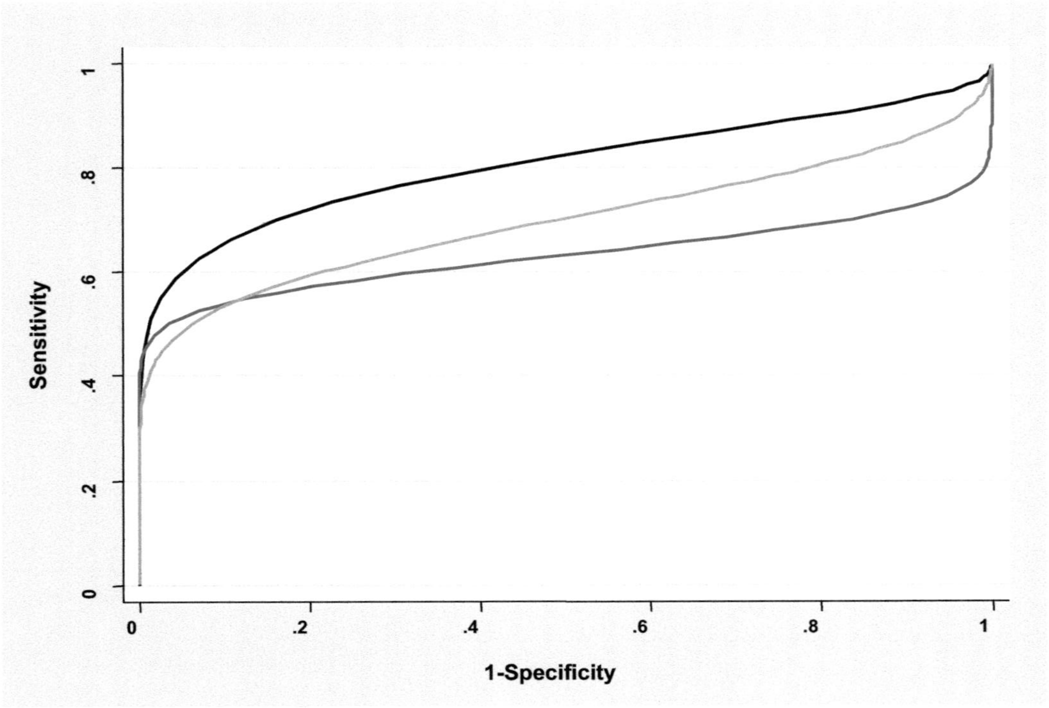

Figure 9. Calculated ROC curves for SICCT (black), SENTRY 100 (dark gray) and GENios Pro (light gray).

Test	AUC	95% CI	S	95% CI	C	95% CI
SICCT (OIE cut-off > 4 mm)	0.80	0.73 - 0.87	51.1%	42.1 - 60.1%	98.6%	97.9 - 99.2%
SICCT (cut-off > 2 mm)	0.80	0.73 - 0.87	66.3%	57.5 - 74.6%	89.2%	86.6 - 91.5%
SENTRY 100 (cut-off ≥ 15 ΔmP)	0.57	0.51 - 0.65	45.5%	39.3 - 52.9%	96.4%	95.4 - 97.4%
GENios Pro (cut-off ≥ 38 ΔmP)	0.64	0.57 - 0.72	47.2%	39.9 - 54.7%	92.4%	90.7 - 93.9%
Meat inspection*	-	-	36.1%	26.6 - 46.9%	90.8%	88.6 - 92.5%
Direct microscopy*	-	-	90.0%	74.4 - 96.5%	66.7%	55.2 - 76.5%
Culture and microscopy*	-	-	93.3%	78.6 - 98.2%	69.4%	58.0 - 78.8%
PCR*	-	-	71.4%	52.9 - 84.7%	100.0%	85.1 - 100%
True prevalence	8.4%	6.1 - 11.0%				

* Estimates are based on modelled latent disease state of the animals and refer to the sample; 95% CI are Wilson confidence intervals

Table 15. Parameter estimates for different diagnostic tests based on results from model 2A. AUC: Area under the ROC curve; CI: confidence interval; S: sensitivity; C: specificity.

	Univariate model			Multiple model*		
Explanatory variable	OR	95% CI	p	OR	95% CI	p
Age	1.15	1.05 - 1.26	<0.01	1.14	1.02 - 1.29	<0.05
Sex	1.59	0.96 - 2.64	0.07	1.11	0.61 - 2.01	0.74
Breed	1.28	0.79 - 2.06	0.31	1.54	0.94 -2.54	0.09
Body condition:						
good	1.00	-	-	1.00	-	-
bad	1.07	0.66 - 1.73	0.79	0.96	0.58 - 1.58	0.86
very bad	2.81	1.33 - 5.95	<0.01	1.96	0.88 - 4.38	0.10

*Model was adjusted for age, sex, breed and body condition

Table 16. Logistic regression for modeled *M. bovis* infections.

	N	%	RR	Fisher
Animals with lesions	**108**	**100%**	**N/A**	**N/A**
Lymphnode lesions	***98***	***91%***	***0.66***	***0.46***
Prescapular lymph nodes	64	59%	1.19	0.67
Mammary lymph nodes	37	34%	1.11	0.82
Head associated	8	7%	0.43	0.44
Popliteal lymph nodes	1	1%	0.00	1.00
Organ lesions	***22***	***20%***	***2.99***	***<0.01***
Lung	17	16%	3.10	<0.01
Liver	8	7%	2.50	<0.04
Others	3	3%	2.50	0.19

Table 17. Lesion distribution and association between lesion localization and modeled *M. bovis* infection. N: Number of animals with a specific lesion localization. %: Percentage of animals with lesions with a specific lesion localization. RR: Risk ratio for modeled *M. bovis* infection. Fisher: Fisher's exact test p-value.

Discussion

Practicability and significance of Bayesian ROC estimation

The performance of diagnostic tests is often setting dependent [249]. Thus, evaluations of diagnostic tests for a given region are a prerequisite for the implementation of local disease surveillance schemes and control measures [249]. However, to date, only few studies have assessed the performance of tests for the diagnosis of BTB in high incidence countries in Africa. Furthermore, test evaluation is hampered by the absence of a gold standard method for the identification of the animal's true disease status. Here, we applied a Bayesian approach for the evaluation of multiple tests for the diagnosis of BTB in a naturally infected slaughterhouse population of cattle in southern Chad. Our approach did not require knowledge of the true disease status of the tested animals. Moreover, it allowed the estimation of the true disease prevalence in the sampled population, the calculation of the BTB disease status of all sampled animals and the evaluation of four post-mortem tests for the diagnosis of BTB.

We have previously reported the evaluation of SICCT, SENTRY 100 and GENios Pro using a subset of the same data [59]. In a gold standard approach, PCR confirmed MTBC infected animals were defined as the positive population and lesion negative animals as the negative population and used for the construction of ROC curves for each test. Drawbacks of this approach were the relatively small amount of confirmed infections and the unknown true disease status of lesion negative animals. Table 18 compares the results from the present and our previously published study [59]. The accordance of our results using the two different approaches further supports the accuracy of our estimates and the practicability of our Bayesian method. Noteworthy, Bayesian modeling gave rise to parameter estimates with in many cases considerably smaller confidence intervals compared to the gold standard approach (table 18).

SICCT

Our results indicated that the most appropriate cut-off for positive SICCT test interpretation was significantly lower then the OIE suggested standard cut-off (> 2 mm versus > 4 mm). However, our criteria for cut-off selection attributed equal weights to sensitivity and specificity and did not consider the disease prevalence and the cost of misclassifications. As an alternative approach for cut-off selection, the misclassification-cost term (MCT) can be calculated for each point of the ROC curve. The point with the lowest MCT value would then be most appropriate for positive test interpretation [243]. This method requires to quantify the cost of false negative (CFN) and false positive (CFP) diagnosis, which we were not able to accurately do. However, the cost of a false negative diagnosis is likely to exceed the cost of a false positive result by several folds as disease transmission amplifies the total economical losses due to BTB. We found that, assuming a disease prevalence of 8.4% (10.0%), a cut-off > 2 mm would be ideal if CFN/CFP lies between 8 and 16 (7 and 13). This suggests that our chosen cut-off values may be acceptable for a broad range of reasonable CFN/CFP ratios.

A cut-off > 2 mm was also found to be most appropriate for positive SICCT test interpretation in a recent study in Ethiopia [65] and in SICCT reactor prevalence studies in Uganda and Tanzania, lower cut-offs than the OIE standard cut-off have been used, however without detailed justification [33,36]. Accordingly, our results are likely to apply for many other countries in sub-Saharan Africa with similar environmental and economic conditions.

SICCT showed a relatively low sensitivity irrespective of whether our suggested or the OIE cut-off was used (table 15). Comparable results were obtained in previous studies in Ireland and Madagascar [87,162]. This relatively weak performance may be explained by several factors. A high proportion of pre-allergic animals at an early stage of BTB infection or a high amount of SICCT anergic animals at a very late disease stage could have accounted for this observation [162]. Antigens of co-infecting NTM strains, cross reacting with PPD-A could also cause false negative test results as well as nutritional stress or concurrent infections with pathogens leading to immuno-depression [162]. For SICCT anergy due to generalized BTB, one would expect the presence of gross visible lesions. Amongst all animals with a modeled *M. bovis* infection and visible lesions (N = 30), 9 or 19 (30% or 63%) did not show a positive reaction to SICCT depending on whether a cut-off > 2 mm or > 4 mm was applied, respectively. This indicates a considerable proportion of SICCT anergic animals (9 or 19 of altogether 83 animals with modeled *M. bovis* infection). Unfortunately, our sample size was too small to conclusively assess the ability of the FPA tests to detect such animals.

Cause of lesions

Our data suggests that a surprisingly high proportion of lesions detected during standard meat inspection at the Sarh abattoir in southern Chad was caused by other bacteria than *M. bovis*. For 72% of the animals in which lesions have been detected, no *M. bovis* infection was modeled. This finding was in line with the relatively low amount of MTBC strains detected in animals with lesions (20 of altogether 108 animals with lesions; table 13). Interestingly, modeled *M. bovis* infection was only significantly associated with organ lesions in general and the presence of lung and liver lesions in particular (table 17). The presence of lymph node lesions was not associated with modeled *M. bovis* infection (table 17). Altogether, this suggests that a significant amount of gross visible lesions detected during standard meat inspection at the Sarh abattoir has been caused by other pathogens than *M. bovis* and that especially a large proportion of the detected lymph node lesions may have been caused by these pathogens.

NTM infections without concomitant *M. bovis* infections have been isolated from 10 out of 50 animals tested by PCR. This could indicate that some of the lesions may have been associated with NTM. This is also supported by the comparatively low specificity of Ziehl-Neelsen staining and microscopic examination of extracted lesions or bacterial cultures in our setting compared to previous studies (table 15) [250-254]. Nevertheless, the low amount of cultures in which AFB have been detected (50 of 108 animals with lesions) suggests that in addition, other pathogens may have been responsible for the detected lesions.

Altogether, our data indicates that the amount of gross visible granulomatous lesions caused by other pathogens than *M. bovis* may be greatly underestimated in this setting. Low recovery of *M. bovis* from cultures of granulomatous lesions have been reported in several studies on BTB in sub-Saharan Africa [43,47,62,67]. It is conceivable that in many of these cases, lesions may have been caused by other pathogens and that these bacteria may have remained undetected e.g. due to the decontamination procedure or different culture growth requirements.

However, it has to be noted that the proportion of lesions due to other pathogens than *M. bovis* is dependent on the accuracy of the meat inspection. Inaccurate meat inspection e.g. biased toward superficial lymph nodes could have distorted the relative proportion of lesions found in different organs. In particular, it is surprising that no lesions were detected in the bronchial or mediastinal lymph nodes, as these are usually the most often affected tissues in bovine tuberculosis [63,64,255,256]. Also, the sensitivity of meat inspection to detect *M. bovis* infected cattle was lower in our setting compared to the results of previous studies [63,64,245,246]. Therefore, the proportion of lesions caused by other pathogens than *M. bovis* may have to be interpreted with caution.

Risk factors

In a previous study on BTB in Chadian cattle we have reported that the prevalence of BTB was significantly higher in Mbororo zebus than in Arab zebus [43]. Our results from the logistic regression analysis could not show any evidence that *M. bovis* infection was significantly associated with breed (table 16). Nevertheless, the presence of lesions was still significantly associated with Mbororo zebus ($N = 944$, $\chi 2 = 5.23$, $p = 0.02$). This observation could suggest that Mbororo breeds in fact, are not more likely to be infected with *M. bovis* but more often develop advanced stages of the disease. Host genetic factors as well as environmental factors or animal husbandry could account for this observation.

Conclusions

In summary, the present study shows the practicability of a Bayesian method for the evaluation of multiple tests for the diagnosis of BTB in naturally infected cattle and in absence of knowledge of the true disease status of the animals. Our model allowed us to compute the disease status of each sampled animal and the modeling results supported our previous observation that the cut-off for positive SICCT interpretation should be lowered to > 2 mm in many countries of sub-Saharan Africa. Moreover, we provide evidence that an unexpectedly high proportion of BTB suspect lesions detected during slaughterhouse meat inspection was due to other pathogens than *M. bovis*.

	SICCT		SENTRY 100	GENios Pro
Cut-off	> 4mm	> 2mm	≥ 15	≥ 38
Bayesian method:				
Sensitivity	51.1% (42.1 - 60.1%)	66.3% (57.5 - 74.6%)	45.5% (39.3 - 52.9%)	47.2% (39.9 - 54.7%)
Specificity	98.6% (97.9 - 99.2%)	89.2% (86.6 - 91.5%)	96.4% (95.4 - 97.4%)	92.4% (90.7 - 93.9%)
AUC	0.80 (0.73 - 0.87)		0.57 (0.51 - 0.65)	0.64 (0.57 - 0.72)
Gold standard approach:				
Sensitivity	20.0% (5.7 - 43.7%)*	65.0% (43.3 - 81.9%)	30.0% (14.5 - 51.9%)	50.0% (29.9 - 70.1%)
Specificity	93.1% (91.1 - 94.6%)	86.7% (84.2 - 88.9%)	94.4% (92.7 - 95.8%)	88.4% (86.1 - 90.4%)
AUC	0.80 (0.71 - 0.88)		0.70 (0.58 - 0.82)	0.67 (0.52 - 0.82)

* 95% binomial exact confidence intervals are indicated if (estimated value) x (sample size) ≤ 5, otherwise Wilson confidence intervals are shown for estimates of the gold standard approach

Table 18. Comparison of parameter estimates derived from the Bayesian model considering no gold standard and from a previously employed model (also see last chapter) assuming the PCR confirmed infections (positives) and the absence of lesions (negatives) as gold standard.

Materials and Methods

Animals

The animal population subjected to this study has previously been described, in detail [59]. A total of 954 slaughter animals were sampled during three intervals of approximately one month between July and November 2005 at abattoirs in southern Chad. We can assume that the tested animals constitute a representative sample of slaughter cattle from a large number of different herds and a big area in southern Chad [59]. Presumably, none of the animals has ever undergone tuberculin skin testing. Four types of phenotypic zebu breeds were encountered: Arab (N = 658), Mbororo (N = 286), Bogolodjé (N = 7) and cross breeds (N = 3).

Physical examination of animals

All 954 animals were physically examined before slaughter. Body condition was assessed by assigning one of the following three scores: 1 – good body condition, 2 – bad body condition, 3 – very bad body condition [59].

SICCT

Valid SICCT testing results were available for 930 animals. SICCT testing and reading was carried out as explained previously and according to standard protocols [59,163].

Fluorescence polarization assays

Valid SENTRY 100 and GENios Pro FPA results were available for 953 and 954 animals, respectively. The methods have been previously described in detail [59,170].

Meat inspection

After slaughter, all 954 animals underwent meat inspection, which included organ and lymph node palpation, visual inspection and incision of organs and lymph nodes according to standard procedures [240]. However, we were not able to fully exclude potential irregularities during the carcass examinations. Meat inspection was done by local meat inspectors at the abattoirs in southern

Chad. Gross visible lesions were detected in altogether 108 of the 954 sampled animals. Lesion containing tissue specimens from all visibly affected organs and lymph nodes were collected and transported on ice to the Chadian National Veterinary and Animal Husbandry Laboratory (Laboratoire de Recherches Vétérinaries et Zootéchniques de Farcha) in N'Djaména and stored at -20°C.

Direct microscopy

Specimens from all 108 animals with lesions were subjected to direct microscopy and processed as previously described [43]. After homogenisation, specimens were colorized by Ziehl-Neelsen staining and examined under the light-microscope for the presence of Acid-Fast Bacilli (AFB). The samples were decontaminated with N-acetyl-L-cysteine sodium hydroxide (0.5% NALC 2% NaOH) and again examined for the presence of AFB under the microscope. If either of the two microscopic examinations revealed presence of AFB the result was considered to be positive.

Culture and microscopy

Specimens of lesions from altogether 102 animals were subjected to culture and microscopy. Decontaminated samples were inoculated into two Middlebrook 7H9 medium flasks containing OADC and PANTA and either glycerol (0.75%) or pyruvate (0.6%) [59]. Samples were put into culture until growth was detected or for a minimum of 8 weeks. Presence of AFB in cultures was examined by Ziehl-Neelsen staining and microscopy [43]. Bacterial growth was detected in cultures of 102 animals; cultures of 50 animals showed presence of AFB by Ziehl-Neelsen staining.

Real-time PCR

AFB containing cultures from 50 animals were subjected to molecular characterization. Heat inactivation of the cultures was carried out as previously explained [59]. Thermolysates were shipped to the Swiss Reference Centre for Mycobacteria, DNA was extracted by means of the InstaGeneTM Matrix (Bio-Rad) and identification of MTBC and NTM strains was carried out by means of Light Cycler® PCR as previously described by Lachnik et al. [241].

Statistical analyses

A Bayesian model was developed to estimate the true *M. bovis* infection status of all sampled animals. The model combined the results of the continuous as well as the binary diagnostic tests for BTB (SICCT, SENTRY 100, GENios Pro, meat inspection, direct microscopy, culture and microscopy and PCR), applied to the same animal population without considering a gold standard. It also allowed the estimation of the true disease prevalence in the sampled population as well as the sensitivities and specificities of the diagnostic tests. A mathematical description of the model and the WinBUGS code are provided in appendix 4 and 5, respectively.

Risk factors for the modeled *M. bovis* infection status were identified by univariate and multiple logistic regression analysis in Stata (Stata/IC v10.0). Association between lesion localisation and modeled *M. bovis* infection was assessed by the Fisher's exact test in Stata.

Bayesian modeling of disease status

We assumed that the distribution of the test values of SICCT, SENTRY 100 and GENios Pro was trivarite normal with means and variance-covariances separately estimated for the diseased and the non-diseased animals. The normality assumption was verified via the shape of the histogram of the test values. In an initial model we have included the data from the multiple post mortem tests for the detection of *M. bovis* infected animals (meat inspection, direct microscopy, culture and microscopy and PCR) and tried to directly model their sensitivities and specificities (models 1A and 1B, table 14). Because the parameter estimations for the binary post-mortem tests were highly sensitive to the prior assumptions [63,64,245,246,250-254,257,258], we eventually excluded the respective data from the Bayesian model formulation (models 2A and 2B, table 14). In order to estimate the performance of the binary tests, we used the modeled latent *M. bovis* infection status of each animal to calculate sensitivities and specificities of the respective tests. Model fit was done in the statistical package WinBUGS (Imperial College and Medical Research Council, UK). The mathematical description of the model and the WinBUGS code are shown in appendix 4 and 5. Real-time PCR was assumed to be 100% specific and animals with a positive PCR test outcome were therefore defined as MTBC infected in all Bayesian models.

ROC curve and cut-off selection

From the estimates of the means and variance-covariances of the multivariate normally distributed continuous test values of SICCT, SENTRY 100 and GENios Pro for the diseased and non-diseased animals, a ROC curve was calculated in Stata. Pairs of 1-specificity and sensitivity were calculated and plotted for all possible cut-off points according to the following formula:

$$\left[1-\Phi\left(\frac{c-\mu_k^{d=0}}{\sqrt{\sigma_{kk}^{d=0}}}\right),\ 1-\Phi\left(\frac{c-\mu_k^{d=1}}{\sqrt{\sigma_{kk}^{d=1}}}\right)\right]$$

Φ is the cumulative distribution of a standard normal variable; c is the cut-off value; $\mu_k^{d=0}$ and $\mu_k^{d=1}$ are the means of the multivariate normal distribution of the test values for the non-diseased (d = 0) and the diseased (d = 1) animal population subjected to test k (k = 1 for SICCT, k = 2 for SENTRY 100, k = 3 for GENios Pro), respectively; $\sigma_{kk}^{d=0}$ and $\sigma_{kk}^{d=1}$ are the variance of the non-diseased and the diseased animal population subjected to test k, respectively.

We considered the point of the ROC plot with the greatest distance from the diagonal line (sensitivity = 1 - specificity) as the best cut-off; this corresponds to the point with the largest Youden index (J = sensitivity + specificity - 1) [242,243]. In cases where several points showed the same distance, the point with the highest sensitivity was chosen. For cut-off selection using the misclassification-cost term (MCT), the point with the smallest MCT value [MCT = (CFN / CFP) P (1 - Se) + (1 - P) (1 - Sp)] was chosen, with CFN and CFP being the cost of false negative and false positive diagnosis, respectively and P being the disease prevalence in the target population [243]. We were unable to accurately estimate CFN/CFP but the cost of false-negative diagnosis is likely to exceed the cost of false positive diagnosis. Therefore, MCT values for each possible cut-off point and different ratios of CFN/CFP were calculated and compared, assuming a disease prevalence of 8.4%, as estimated by our model. In addition, MCT values for different CFN/CFP ratios were calculated for a 10.0% disease prevalence.

Acknowledgements

We would like to thank Prof. Erik C. Böttger, Dr. Boris Böddinghaus, Dr. Burkhard Springer and the technicians of the Swiss National Centre for Mycobacteria in Zurich for providing technical support and laboratory facilities. We are indebted to the cattle holders, traders and butchers who were collaborating with us within this project. Our work has received financial support from the Swiss National Science Foundation (project no. 107559).

Part IV

General discussion and conclusions

Considerations for the control and surveillance of bovine tuberculosis in sub-Saharan Africa

Abstract

Livestock producing countries of the developing world bear most of the global burden of bovine tuberculosis (BTB). The disease is present virtually in whole of Africa with only few countries being able to apply appropriate control measures. Surveillance is generally based on slaughterhouse meat inspection but some countries have implemented occasional active surveillance in herds by means of the tuberculin skin test. However, recent results suggest that the OIE standard cut-off for tuberculin test interpretation may not be appropriate for all countries of sub-Saharan Africa. The majority of the cattle in Africa are raised in a traditional extensive livestock production system. The characteristics of the local cattle husbandry along with economic, political and historical factors and the particular natural environment, have to a large extend influenced the distribution and prevalence of BTB. Distinct clonal complexes of *Mycobacterium bovis* are localized to different regions of the continent and several studies in different parts of the world including Africa suggest that *M. bovis* was originally imported from Europe. There is accumulating evidence for a particular involvement of non-tuberculous mycobacteria in lesion formation in African cattle and possibly wildlife. These characteristics and the lack of financial resources have important implications with regards to BTB surveillance and control. This review summarizes our research results on the molecular epidemiology and diagnosis of BTB in Africa and discusses our findings in a broader context by considering the specific African background. Moreover, on the basis of the currently available information, potential, locally adapted control and intervention strategies shall be identified.

Introduction

Bovine tuberculosis (BTB) is caused by *Mycobacterium bovis* and constitutes an economically important disease, primarily affecting cattle. It influences trade and can lead to considerable losses in meat and milk production [19]. Moreover, *M. bovis* infections have been detected in wildlife, where the disease can have severe consequences for the ecosystem [20,22]. *M. bovis* is also of public health concern as the bacteria can be transmitted to humans most notably through close contact with infected animals and consumption of raw milk [23]. BTB is considered a neglected zoonosis and closely related to poverty [26]. In Africa, the disease is present on virtually the whole continent [5,32,45]; however, only few countries report to apply specific control measures [32]. Data on disease prevalence is generally scarce due to the lack of disease surveillance schemes and laboratories with the capacity to culture and accurately characterize mycobacterial strains. As a matter of fact, many laboratories in Africa rely solely on Ziehl-Neelsen staining and microscopy for the diagnosis of tuberculosis infections [32]. Consequently, little information on the zoonotic transmission of *M. bovis* exists, and the frequency of *M. bovis* infections in humans may be underestimated [45].

Control measures against BTB in Europe have been based on test and slaughter strategies. Animals have been tested using the intra-dermal tuberculin skin test; nowadays, the Bovigam® test (Prionics) is increasingly used as an ancillary test [162]. Positively tested cattle have been subjected to slaughter. In addition, movement restrictions have been implemented [194,259] and cattle carcasses have been regularly inspected by veterinary officers at abattoirs [260]. Carcasses exhibiting gross visible lesions have been fully or partially condemned according to the severity and dissemination of the lesions. Moreover, farmers have been compensated for the accruing losses.

In Africa, BTB control in livestock is mostly limited to slaughterhouse meat inspections. In some countries occasional tuberculin skin testing of animals from intensive dairy farms is performed; however, attempts to evaluate diagnostic tests for BTB in naturally infected cattle in Africa are scarce. Moreover, existing control measures are often not well implemented and farmers do not receive compensation for culled livestock [45].

In the this review, the results from our studies on the population structure of *M. bovis* in Africa and the evaluation of multiple diagnostic tests for BTB in African cattle shall be recapitulated and discussed in a broader context. Moreover, the specific characteristics of BTB in Africa shall be considered to give recommendations for possible adapted interventions and control programs.

Ante-mortem diagnosis of bovine tuberculosis

Control measures against BTB must primarily target the transmission of *M. bovis* within and between the cattle herds; moreover, wildlife populations can serve as reservoirs of *M. bovis* and have been shown to play an important role in some African countries [20,79]. Surveillance aims at early detection of infected animals in order to interrupt the transmission of pathogens. However, surveillance is highly dependent on the diagnostic tests available and different tests may be appropriate in different settings.

A new cut-off for the tuberculin skin test in sub-Saharan Africa

The single intra-dermal comparative cervical tuberculin (SICCT) test is, although imperfect, currently the most widely-used method for the diagnosis of BTB [162]. Purified protein derivatives of *M. bovis* cultures (PPD-B) and *M. avium* cultures (PPD-A) are intra-dermally injected and increase in skin-fold thickness is measured 72 h after. According to OIE guidelines, the test is positive if the difference between the increase in skin thickness at the site of PPD-B and PPD-A injection is more than 4 mm (> 4 mm), inconclusive if between 1 and 4 mm and negative if zero or less [163]. However, receiver operating characteristics (ROC) analyses of SICCT testing in Chadian cattle suggested that SICCT would perform considerably better if the cut-off value was lowered to > 2 mm (see part III of this thesis: Diagnosis of bovine tuberculosis in Chadian cattle, [59,60]). The same result was obtained if ROC analysis was based on a gold standard definition with confirmed *Mycobacterium tuberculosis* complex (MTBC) infected animals as the positive population and lesion negative animals as the negative population or if a Bayesian approach in absence of a gold standard was applied [59,60]. In another study in Ethiopian cattle, similar results were obtained [65] and in SICCT reactor prevalence studies in Uganda and Tanzania, lower cut-offs than the OIE standard cut-off have been used, although without detailed justification [33,36]. Standard SICCT cut-offs should be revised for sub-Saharan Africa as a cut-off > 2 mm may be more appropriate in many areas of the continent with similar environmental and economical conditions.

Searching the right test

SICCT testing is based on the early Th1 type cell mediated immune (CMI) response and detects the allergic reaction to intra-dermally injected *M. bovis* antigens. Bovigam® is another CMI response based test, which detects and quantifies the production of IFN-γ in whole blood cultures exposed to bovine and avian tuberculin (PPD-B and PPD-A) [162,164]. The Bovigam® test shows some practical advantages compared to SICCT as the test does not directly interfere with the host's immune response and does not affect subsequent retesting of the animals; moreover, revisiting of the animals for test interpretation is not necessary with Bovigam® [162,164]. However, a disadvantage of Bovigam® is the requirement that collected blood samples have to be processed relatively quickly after collection and a good laboratory infrastructure is needed [162,164]. The necessity of fast processing of blood samples for Bovigam® and the need to revisit cattle herds for test evaluation of SICCT makes both tests relatively impractical for the testing of often transhumant and nomadic herds in rural Africa.

SICCT and Bovigam® are both based on the CMI response [165]. However, it was shown that CMI responses can wane as BTB disease progresses and animals can become anergic; this can result in false negative SICCT and Bovigam® test results [162]. The problem is exacerbated by a possibly higher infectivity of late stage diseased animals, which are thought to shed higher amounts of infectious bacteria [165]. In low income countries, where control measures are absent, the predicted higher prevalence of such animals might considerably affect disease spread and persistence [162,165,168]. Indeed, our evaluation of multiple diagnostic tests for BTB indicated that a considerable amount of SICCT anergic *M. bovis* infected cattle may be present in Chad [59,60] and very likely, this is also the case in many other African countries.

While CMI response is waning towards late disease stages, humoral immune response is increasing [162,165]. Hence, serological tests, detecting the presence of *M. bovis* specific antibodies might be more appropriate to detect animals at a late stage of BTB. We evaluated two distinct fluorescence polarization assays (FPA), which were based on the detection of antibodies against MPB70 protein, for disease diagnosis in Chadian cattle [59,60]. Although the number of confirmed MTBC infected animals was too small to allow any statistically significant conclusion, there was no indication that anergic animals were more likely to be detected by FPA (see part III of this thesis, [59,60]).

Our results suggested that SICCT performed generally better than FPA in its current form. For the Chadian setting, we have estimated the sensitivity and specificity of SICCT at 66% and 89%, respectively, when our suggested cut-off > 2 mm was applied [59,60]. This performance is still very unsatisfactory and raises the question of which diagnostic test should be used in sub-Saharan Africa. Ideally, the test should be highly sensitive, easy to apply and possibly a serological test, which also allows the detection of late stage diseased animals.

We would therefore strongly encourage further research on the development of diagnostic tools for the specific identification of SICCT anergic animals. As already stated above, there is evidence that animals with more progressive forms of BTB may represent the greatest threat for disease spread [165] and our data supports the idea that a considerable proportion of such animals may be present in developing countries.

Molecular epidemiology of *Mycobacterium bovis* infections

Molecular typing of *M. bovis* strains isolated from cattle at abattoirs has helped us elucidate essential characteristics of BTB in Africa and will continue being of great importance for the surveillance of *M. bovis* infections. The sampling- and typing strategies for *M. bovis* genotyping shall therefore be discussed at this point prior to review some of the results that have been obtained from such investigations.

Sampling

The effectiveness of disease surveillance is highly dependent on the applied sampling framework. Surveys on the molecular epidemiology of *M. bovis* infections in cattle in Africa have mostly been based on sampling schemes targeting one or few major abattoirs of the country [43,47,52,61,69,261]. I shall therefore discuss the appropriateness of such an approach in order to assess the regional bacterial population structure of *M. bovis*.

Livestock production constitutes the main agricultural output in the arid and semi-arid zones of East- and West Africa [28]. Cattle from these areas are generally kept in traditional extensive farming systems, which focus on milk production and herd growth (A. Shaw, pers. comm.). Moreover, livestock production in the arid and semi-arid zones of Africa is often characterized by nomadic or transhumant herd movements [28]. The situation differs in other areas of Africa. E.g., in our studies at Algiers and Blida abattoir in Algeria, mostly exotic breeds, generally used for intensive dairy farming, were encountered [261]. However, a common trait of both livestock production systems is their focus on milk production and the fact that only few animals are sold from the same herds at the same time. Moreover, in many cases, the animals are at first sold to different traders, which in turn sell on the animals to different butchers (N. Sahraoui and B.N.R. Ngandolo, pers. comm.; [47,59,261]). It may be in part due to this, but also due to the frequent transhumant livestock movements, that many of the cattle encountered at abattoirs in Africa can originate from a very distant area from the abattoir. Transhumance may play a minor role in Algeria, nevertheless it has been stated that some of the animals encountered at the abattoir of Algiers originated from an approximately 300 km distant region around Sétif (N. Sahraoui, pers. comm.). Taken together, the small number of animals sold from the same herd at the same time, the multiple selling-on of animals through traders and the long distance cattle movements result in an efficient intermixing of animals from different origins and herds at the abattoirs. Thus, we can assume that the sampling from large slaughter houses constitutes an efficient strategy to obtain a random sample of cattle from many herds and an extensive area. However, for the same reasons and because of poor documentation of cattle trade in most African countries, the actual origin of the individual animals cannot be traced. Consequently, molecular epidemiological studies on *M. bovis* strains isolated from slaughter cattle at abattoirs are clearly not suitable for the investigation of the geographical distribution of *M. bovis* genotypes. Sampling from abattoirs may also be biased as the considerable amount of animals subjected to private slaughter is missed. It is important to emphasize that the sampling from a single abattoir will not be sufficient to assess the country-wide bacterial population structure but it may help, in the best case, to detect the most frequent genotypes of *M. bovis* for an extensive area. Short sampling periods may constitute another problem for the full assessment of circulating *M. bovis* strains. We performed two slaughterhouse surveys in Chad, one at

N'Djaména abattoir between 2002 and 2004 and the second in 2005 at Sarh abattoir, located approximately 500 km from N'Djaména [43,59]. Of altogether nine spoligotype patterns detected in *M. bovis* strains from Sarh, five were also present in strains from N'Djaména (appendix 3, [262]). The three most common types in Sarh were also the three most common types in N'Djaména. However, four of nine spoligotype patterns from Sarh and seven of twelve spoligotype patterns from N'Djaména were not detected in strains from the other study site. Based on our results we can conclude, that short period slaughterhouse surveys may give satisfactory results for a first identification of the most prominent *M. bovis* strains circulating in a given area, provided that the slaughterhouse cattle population is likely to originate from many herds and a large area. However, molecular epidemiological studies assessing the geographical distribution of *M. bovis* strains in Africa would gain enormous information if animals could be traced back by e.g. sale or transport certificates, transferred from the animal owner, via all intermediaries to the abattoir.

Meat inspection and culture

Obviously, the correct and systematic execution of the meat inspection at abattoirs is crucial for effective post-mortem surveillance of BTB. However, it goes beyond the scope of this review to discuss the inspection procedures and I shall therefore refer to the specific technical literature that has been published on this subject [240]. The same applies to the protocols for mycobacterial culture for which extensive literature is available [263]. In a project funded by the Wellcome Trust Livestock for Life Initiative we have established an African BTB network of veterinary authorities and research representatives from a number of African countries. We have recently started to compile standard operating procedures for mycobacterial culture and genotyping for this network. The respective protocols shall be made freely available in the near future through the network website (http://www.africa-btb.net/).

Molecular typing

Molecular techniques have replaced the laborious and often erroneous [13,264] biochemical methods for the identification and differentiation of MTBC strains in industrialized countries. Although, biochemical tests are still used in diagnostic laboratories of developing countries, the establishment of molecular techniques in these laboratories is essential. Despite enabling the much faster and unambiguous identification of MTBC strains, they are indispensable for the investigation of the molecular epidemiology of tuberculosis infections. The use of different typing techniques for MTBC has been extensively reviewed [17,125-128]. One of the major activities within our Wellcome Trust funded African BTB network is devoted to the training of laboratory staff in molecular methods for the diagnosis and the molecular epidemiology of *M. bovis* infections. Within this network, we aim at establishing simple PCR based techniques for the identification and differentiation of *Mycobacterium* spp. and MTBC in various African laboratories [14,145,265]. Moreover, we are assisting the establishment of spoligotyping [53] and VNTR typing [1,142] as means for molecular epidemiological investigations.

Spoligotyping has been our method of choice for an initial assessment of *M. bovis* strain diversities in population surveys and for the identification of possible strain families [135,183,184]. Importantly, the construction of extensive spoligotype pattern databases (www.Mbovis.org, [136]) has facilitated the comparison of results from different countries and helps elucidate the distribution and spread of strains. VNTR typing is another simple method for *M. bovis* genotyping with a higher discriminatory power than spoligotyping [1,142]. However, international databases are presently not available and the method has not yet been standardized. This may be in part due to the observation that distinct loci appear to show a suitable discriminatory power in different countries. This can be demonstrated by our following observation: We subjected 88 *M. bovis* strains isolated from Algerian cattle to VNTR analysis targeting ETR loci A-E [138,261] and MIRU 26 and 27 ([139]; unpublished results). High discriminatory power of VNTR loci defined by a calculated allelic diversity > 0.25 [142] was obtained for ETR loci A, B, C and MIRU 27. In our previous study on *M. bovis* strains from Chad, in addition to these loci MIRU 26 was also highly discriminative [142]. Strains isolated from Belgian and Irish cattle also showed a high allelic diversity for MIRU 26 but low diversity for ETR C. Moreover, MIRU 27 did not show any discriminatory capacity in strains from Ireland [131,266].

An important even though infrequently observed drawback of spoligotyping and VNTR typing is the occurrence of homoplasy, that is, the presence of identical typing patterns in unrelated strains [1]. E.g., a very frequent ETR A-F pattern (7 5 5 4* 3 3.1) of strains from Mali belonging to the provisionally named Af5 clonal complex was identical to the pattern of strains isolated from Chadian cattle and belonging to the distinct Af1 clonal complex (appendix 3, [262]). Homoplastic markers are less appropriate for phylogenetic analyses and LSP and SNP analyses have turned out to be more powerful tools for the elucidation of the deep phylogeny and identification of clonal complexes of MTBC strains [1,144].

Population structure of *Mycobacterium bovis* in Africa

Clonal complexes

We have identified a clonal complex of strains of *M. bovis* present at high frequency in cattle from Central-West Africa and we named this group of strains the African 1 (Af1) clonal complex [262]. Moreover, we showed that sub-populations of Af1 are geographically localized to different countries. These findings have been discussed previously [262]; however, here, some aspects of particular importance for BTB control in Africa shall be highlighted.

The large spread of Af1 may be explained by the common long distance transhumant livestock production system in Africa. However, geographical localization of *M. bovis* sub-populations, identifiable through VNTR typing, suggests that trans-border movements of cattle have played a relatively minor role in shaping the present *M. bovis* population structure in the different West African countries. This suggests that BTB control could to some extend be accomplished without the collaboration of neighboring countries. This was not expected and is as such an important finding. However, the role of trans-border cattle migrations for the persistence of BTB may increase once national intervention campaigns would have achieved to reduce the country-wide disease prevalence. This is especially the case if neighboring countries do not apply BTB control measures. In this respect, it is nevertheless worthwhile to aim at coordinated control measures in the whole region delimited by the Af1 clonal complex. Importantly, our results also highlight the value of establishing molecular techniques such as spoligotyping and VNTR typing in order to monitor the importation of extra-national strains.

Our findings and recent work performed at VLA, UK, suggest the presence of other clonal complexes in other parts of Africa. In Mali, the presence of two distinct clonal groups has been identified (Af1 and provisionally named Af5; figure 4; figure 10). We have recently conducted a molecular characterization of strains of *M. bovis* isolated from slaughter cattle at Morogoro abattoir in Tanzania. Our preliminary results also suggested the presence of at least two clonal complexes in Tanzania (figure 11). One group of strains lacking spacers 3-7 (provisionally named Af2) was similar to strains isolated from Ethiopian [66], Ugandan [69] and Burundian (unpublished work at VLA, UK) cattle (figure 10; figure 11). Spoligotype patterns of the second group characteristically lacked spacers 5, 11 and 12. Absence of spacers 5 and 11 was also frequent in strains isolated from cattle in South Africa [86] and unpublished work at VLA suggests a link of this group of strains (provisionally named Eu1) to strains from UK cattle. Possibly, a third clonal complex exists in Tanzania (Af4) but this supposition needs to be verified. However, in summary our results indicate that strains from Tanzanian cattle lacking spacers 5 and 11 are linked to strains from South Africa (Eu1) and strains lacking spacers 3-7 are linked to strains from Uganda, Burundi and Ethiopia (Af2; figure 10).

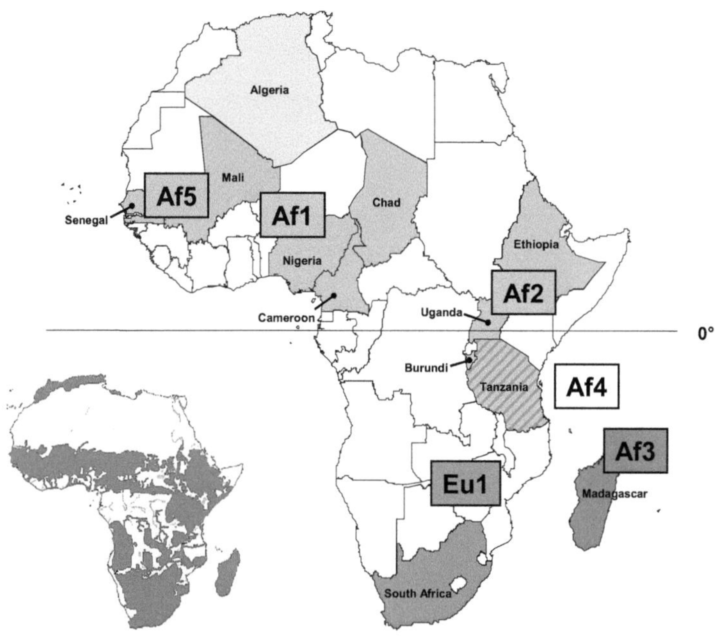

Figure 10. The preliminarily identified clonal complexes in Africa. Strains of Af1 have been found in Cameroon, Chad, Mali and Nigeria. Af2 is present in Ethiopia, Uganda, Burundi and Tanzania. Af3 is present in Madagascar. Based on work done at VLA, another clonal complex, named Af4, is thought to be present in Tanzania. Strains of Af5 have been isolated from cattle in Mali and Senegal. Eu1 is present in the UK, South Africa and Tanzania. A first survey of *M. bovis* strains from Algeria could not identify strains from any of the 6 clonal complexes Af1-5 and Eu1. However, strains from Algerian cattle show a close link to strains from continental Europe. The smaller map shows the cattle distribution in Africa in grey (taken from Hanotte et al. [212])

	SB number	N	%	1	2	3	4	5	6	7	8	9	10	11	12	13	14	15	16	17	18	19	20	21	22	23	24	25	26	27	28	29	30	31	32	33	34	35	36	37	38	VNTR profile
Group I	SB0133	9	60.0%	■	■						■		■	■	■	■	■	■		■	■	■	■	■	■	■	■	■	■	■	■	■	■	■	■	■	■	■	■	■	■	3 2 5 4* 3
	SB1446	1	6.7%	■	■						■		■		■	■	■	■		■	■	■	■	■	■	■	■	■	■	■	■	■	■	■	■	■	■	■	■	■	■	- 2 5 4* 3
	Total	**10**																																								
Group II	SB0425	3	20.0%		■		■		■	■	■		■				■	■		■	■	■	■	■	■	■	■	■	■	■	■							■		■	■	4 5 5 4* 4
	SB0425	1	6.7%		■		■		■	■	■		■				■	■		■	■	■	■	■	■	■	■	■	■	■	■							■		■	■	4 6 5 4* 4
	TANZANIA1	1	6.7%	■	■		■		■	■	■		■			■	■	■		■	■	■	■	■	■	■	■	■	■	■	■	■	■	■	■	■	■	■	■	■	■	4 5 5 4* 4
	Total	**5**																																								

Spacer number

Figure 11: Spoligotypes and VNTR typing patterns of *M. bovis* strains isolated from slaughter cattle at Morogoro abattoir in Tanzania. Spacers 39-43 were absent from all spoligotype patterns. SB numbers were taken from www.Mbovis.org. Pattern TANZANIA1 is not reported in www.Mbovis.org. VNTR typing targeted loci ETR A – E [138].

Strains of Eu1 have not been observed in Uganda, Burundi or Ethiopia although, through Af2, a link between Tanzania and these countries exists. Likewise, Af2 strains have not been detected in South Africa. Af5 strains, although detected in Mali, have not been detected in Nigeria, Chad or Cameroon although, through Af1, a link between these countries and Mali exists. This indicates, that the spread of the four clonal complexes Af1, Af2, Af5, Eu1 may still be continuing. Alternatively, either of the clonal complexes may be replaced by the other more recently emerged group, potentially possessing a selective advantage. However, factors related to livestock distribution and animal husbandry very likely also play an important role with regards to the spreading of the distinct clonal groups. The cattle population density in Africa is concentrated in several areas, which are approximately matching the distribution of the clonal complexes of *M. bovis* strains (figure 10, [28,267]). Important factors influencing livestock distribution in Africa are the cultural value of the animals, availability of food and water and the presence of acute and fatal animal diseases like contagious bovine pleuro-pneumonia (CBPP), rinderpest (although almost completely eliminated), east coast fever and livestock trypanosomiasis [267]. Assuming that environmental conditions have largely influenced the limits of spread for different clonal groups of *M. bovis*, one would expect similar distributions for other cattle diseases in Africa. A recent survey on the global genetic population structure of *Bacillus anthracis* revealed the presence of 12 distinct clonal groups throughout the world [268]. Interestingly, strains from Tanzania and South Africa belonged to the same group but strains from Ethiopia were distinct. Moreover, Maho et al., reported a new genotype of *B. anthracis* strains in Chad, which was not detected in South and East African countries [269]. It is conceivable that environmental factors have contributed in shaping the molecular population structure of *M. bovis* in Africa and that the same factors could have influenced the distribution of genotypes of other cattle pathogens.

Where did Mycobacterium bovis come from?

The predominance of only few clonal groups in Africa could be explained by a selective advantage of these strains. In fact, if Chad, Nigeria and Cameroon had had a prior population of *M. bovis* that was replaced by the Af1 clone it would be difficult to explain, without invoking selection, how Af1 went nearly to fixation in three countries.

Alternatively, the predominance of Af1 in West Africa and the other clonal complexes in other parts of Africa could be explained by the importation of *M. bovis* from outside Africa and the previous absence of *M. bovis* on the continent. Thus, clonal expansion of the different complexes could be a result of geographical opportunity. Furthermore, the observation that strains belonging to the same clonal complexes went nearly to fixation in a number of countries could be due to genetic drift [262]. Indeed, evidence for a French origin of Af1 strains in Cameroon has been discussed by Njanpop-Lafourcade et al., [61] and indications that *M. bovis* has been imported from France (or more generally continental Europe) to Mali and Algeria have been discussed previously (see part II of this thesis: Molecular epidemiology of *Mycobacterium bovis* infections in Africa, [47,261,262]). Likewise, spoligotype patterns similar to patterns of strains from the UK have been found in countries with traditional links to the UK such as Ireland [134], Australia, [195] and Argentina [196]. In fact, it would be relatively easy to propose a link between most of the strains found outside Europe and strains of continental Europe or the UK based on the similarities of the spoligotype patterns. E.g., *M. bovis* isolated from cattle from Brazil [270], Iran [195], the French West Indies and Tunisia [179] show spoligotype patterns identical or similar to patterns from continental European strains and patterns of strains from Madagascar [88] could be derived from European patterns by only one or two spacer deletion events. Exportation of *M. bovis* from Europe to Africa and other continents could be explained by close trade links, especially during the colonial periods [61]. *M. bovis* has only been reported in Africa at the beginning of the 20th century [271] and in line with these considerations, Meyers and Steele (1969) suggested that *M. bovis* has emerged in Europe and has spread from northern Italy to western Europe and the UK. According to Webb (1936), *M. bovis* was then scattered throughout the world through exportation of infected cattle from mainly the UK and the Netherlands to their former colonies [20]. However, these thoughts are still speculative and the evolutionary relationship between European and other strains of *M. bovis* may not be disclosed before sequencing of a number of strains from in and outside Europe allow in-depth genome analyses and the construction of comprehensive phylogenetic trees.

Characteristics of bovine tuberculosis in Africa

Owing to the lack of BTB control schemes and the distinctive livestock production system in most of sub-Saharan Africa, BTB exhibits particular characteristics, which may be uncommon in other regions and which need to be considered for the development of intervention strategies. Some of the specific problems of BTB in Africa have been discussed previously [5,32,45]; however, our and other recent studies revealed new insights into the epidemiology of BTB and are of importance for disease control.

Importance of non-tuberculous mycobacteria

There is accumulating evidence that non-tuberculous mycobacteria (NTM) infections may be responsible for a considerable amount of lesions found in cattle carcasses during standard meat inspection at abattoirs in sub-Saharan Africa.

In previous studies in Chad [58], Uganda [69], Ethiopia [67] and the Sudan [62], a high proportion of NTM strains has been isolated from cattle carcasses exhibiting gross visible lesions. In Chad and Sudan, especially NTM strains of the closely related taxa *M. fortuitum* and *M. farcinogenes* were frequently extracted from cattle lesions. Bayesian modeling of the *M. bovis* infection status of slaughter cattle sampled at the Sarh abattoir in southern Chad allowed us to estimate the proportion of animals with lesions caused by other bacteria than *M. bovis* at more than 70% [60]. From one third of the animals confirmed to be infected with *Mycobacterium* spp., NTM were isolated without any apparent co-infecting strains of *M. bovis* [60]. This could indicate that some of the lesions may have been caused by NTM. Noteworthy, prior to our studies on BTB in Chad, it was believed that bovine farcy caused by *M. farcinogenes*, was the primary agent responsible for lesions detected in Chadian cattle carcasses (C. Diguimbaye-Djaïbe, PhD thesis). However, in addition, other pathogens than *Mycobacterium* spp. may have been responsible for some of the lesions found in slaughter cattle from southern Chad. Interestingly, our Bayesian analysis revealed that modeled *M. bovis* infection was only associated with organ lesions but not with lymph node lesions (table 17, [60]). This could indicate that other pathogens than MTBC strains may have in particular accounted for lesions detected in the lymph node tissues. Similarly, our study at the Bamako abattoir in Mali provided evidence that *M. bovis* was primarily associated with the presence of lung lesions and that other pathogens may have accounted for lesion formation in organs outside the lung [47]. Preliminary surveys in wildlife populations from Tanzania and Ethiopia (unpublished results) also suggested a high prevalence of NTM infections. However, isolation of NTM species from animals exhibiting lesions during slaughterhouse meat inspection is rare in France (M. Boschiroli, per. comm.) and was uncommon in our study in Algeria [261]. Altogether, these results suggest a particular involvement of other bacteria than *M. bovis* and in particular NTM strains in lesion formation in cattle of sub-Saharan Africa.

Exposure to NTM strains has been implicated in the induction of a low-level protection from *M. bovis* infection [272] but can also cause interferences in disease diagnosis [272-274] and affect the BCG protection efficacy [275] due to certain cross reacting antigens. The possible high frequency of NTM infections in cattle of sub-Saharan Africa is therefore of considerable relevance with regards to BTB surveillance and intervention programs.

Late stage diseased animals

The importance of SICCT anergic and late stage diseased animals with regards to diagnosis of BTB has been discussed before. A high frequency of SICCT anergic animals in Africa could be explained by the absence of control measures and by the locally predominant livestock production system. Because of the focus on milk production, mainly young males and old cows are slaughtered. Older cows are only slaughtered when milk production decreases or when the animals become too weak to continue herd migration. Therefore, in absence of any regular BTB control schemes, disease can progress without restraint. Also, because late stage diseased animals are thought to be more infective [165], old cows are likely to contribute more to disease spread than younger animals.

Interestingly, in our study in Chad, the mean age of the male cattle encountered at the abattoir of Sarh was rather low (3.4 years; CI: 3.1 – 3.6 years) although comparable with the situation in other livestock systems in West Africa [276]. The selling of relatively young males may indicate herd de-stocking, because of poor grazing in this area or other problems (A. Shaw, pers. comm.).

Transmission

The primary site of *M. bovis* infection in cattle is the respiratory tract; this is also reflected by the predominant occurrence of lesions in the lung and associated lymph nodes. *M. bovis* is therefore mainly transmitted by inhalation of infectious aerosol droplets. In intensive ("landless" [28]) livestock production systems, the close contact of animals in a confined area promotes transmission of the infectious agent; BTB prevalence rates are therefore generally higher in intensive farming systems than in extensive free-range production systems [5,31,32,93]. In our study at Bamako abattoir in Mali, the apparent lesion prevalence was 1.8% [47] and significantly lower than the SICCT reactor prevalence previously measured in 36 cattle herds from the peri-urban region of Bamako (18.6%; [46]). One explanation for this discrepancy could be a high abundance of early stage infected animals, which do not show visible lesions. However, given the fact that an extensive amount of animals encountered at the abattoir of Bamako originated from the pastoral area, a more likely explanation is the lower prevalence of BTB in these animals compared to animals from the peri-urban region of Bamako.

In transhumant and nomadic livestock production systems, disease transmission is thought to occur at places where close contact of animals is frequent, such as water points. In Africa, grazing animals usually gather at night for protection from predators; moreover, during daytime, they often assemble under trees or other shaded areas [5]. A recent study from Ethiopia reported a distinct lesion distribution in cattle grazing outside in pastures compared to cattle kept indoors [31]. Mesenteric lymph nodes were significantly more often affected in cattle that were kept outdoors, indicating that pathogen ingestion may be an important transmission route in pastoral animals [31]. In this respect, it may be conceivable that pasture gets contaminated through e.g. fecal excretion of *M. bovis*. Moreover, it has been shown that *M. bovis* can survive in feces for up to four weeks [277]. An important route of transmission however, may also be the very early infection of suckling calves with *M. bovis* infected milk. In our study in southern Chad, a high proportion of young animals reacted positively to SICCT and the mammary lymph nodes were the second most affected tissues with lesions (table 9). Young animals may remain latently infected until they become themselves lactating and develop active tuberculosis. An emerging working hypothesis could be that under extensive pastoral conditions animal to animal transmission may take place to a high extend by the ingestion of infected milk and less by environmental contamination and aerosol transmission.

Customs or cultural practices can also influence the transmission of *M. bovis* in cattle. E.g., in Chad, animals are often exposed to smoke from fire places in order to protect the cattle from mosquito bites. However, smoke can affect the nasal mucosa of the animals, rendering them more susceptible to infection with airborne infectious aerosol.

Influence of host genetics

The influence of host genetics on tuberculosis disease is currently extensively investigated for human tuberculosis. Some host gene polymorphisms have been associated with the development of diseases such as polymorphisms in the solute carrier family 11, member 1 (SLC11A1, formerly NRAMP1) [278,279] and recent findings suggest a variable host-pathogen compatibility of distinct human ethnic groups and different strains of *M. tuberculosis* [150].

There is evidence for a differential susceptibility to tuberculosis between exotic and local cattle breeds in Africa [32]. E.g., in a recent study in Ethiopia, BTB prevalence in exotic Holstein cattle was significantly higher than in zebu or cross-breeds [41]. Unpublished results from our study in Algeria also suggested a higher prevalence of *M. bovis* infections in exotic animals compared to indigenous animals and cross-breeds. But differences in susceptibility may also exist between different local zebu breeds. In Chad, prevalence of gross visible lesions and BTB was significantly higher in Mbororo than in Arab zebu breeds [43,59]. However, Bayesian modeling of the infection status of slaughter cattle from southern Chad could not show any evidence that modeled *M. bovis* infection was significantly associated with breed [60]. Altogether, this could suggest that Mbororo breeds may in fact not more likely become infected with *M. bovis* but more often develop advanced stages of the disease [60].

However, it is a difficulty to prove that observed discrepancies in disease prevalence are due to inherent genetic factors of different cattle breeds. Instead, the observation could be confounded by a differential breed specific animal husbandry or by environmental influences. In our study in Chad,

although Arab and Mbororo cattle have been frequently intermixed in the same herds regardless of the cattle owners' ethnicity, we have observed a significantly different sex ratio between the two breeds at the abattoirs [59]. This could indicate different livestock management strategies for different breeds. At the shore of lake Chad, Mbororo animals were observed to be more closely situated to the water than Arab breeds. Although highly speculative, this could e.g. expose Mbororo animals more often to waterborne parasitic diseases (e.g. *Fasciola* spp.), which in turn could enhance their susceptibility to BTB. Interestingly, despite distinct phenotypic characteristics, analysis of the genetic relatedness of both breeds using a set of 21 microsatellite loci could not reveal any significant difference between Mbororo and Arab cattle (appendix 1).
Identification of BTB resistant breeds or cattle husbandry practices that may account for increased susceptibility to BTB would be of great value. Therefore, further research on the influence of host genetic factors as well as animal husbandry practices of different ethnic groups of pastoralists, is highly needed to identify drivers of susceptibility in BTB. Moreover, differential susceptibility in animal diseases in Africa should always be considered from the reality of multi parasitism.

Implications for bovine tuberculosis control in Africa

The particular characteristics of BTB in sub-Saharan Africa must be taken into account for considering potential control options. Due to several factors, and above all the general lack of financial resources, conventional test and slaughter schemes may not be feasible or ill suited for Africa. Therefore, new approaches for BTB control should be considered.
From a public health perspective, the first priority with regards to BTB control is the prevention of zoonotic transmission of *M. bovis* from animals to humans. Next, the control of the disease in cattle needs to be addressed. BTB is primarily of economic concern and infections in domestic cattle may account for the majority of the disease burden. Finally, *M. bovis* infections have to be controlled in wildlife populations. Although the importance of wildlife infections in Africa has not yet been sufficiently investigated, it is well-known that wildlife reservoirs can pose great difficulties to BTB eradication programs [20,22]. Moreover, infections in wildlife can have considerable consequences for the

ecosystem, which are difficult to predict and can indirectly affect many areas of private and economic interest.

Prevention of zoonotic transmission

Although eradication of BTB in cattle will also decrease the prevalence of *M. bovis* infections in humans, transmission may be more directly and cost-effectively prevented by implementation of regular milk pasteurization schemes. In the USA and probably in most other industrialized countries, the public health benefits obtained from BTB control programs were primarily a result of milk pasteurization and to a lesser extend due to the eradication of BTB in cattle [89]. Abattoir meat inspections can play another important role for the prevention of zoonotic *M. bovis* transmission if consumption of raw or undercooked meat is practiced. In Africa, implementation of both control measures could be relatively easy and inexpensive. Importantly, such interventions would not only target *M. bovis* infections but also transmission of other zoonotic diseases such as Brucellosis and Q-fever. Moreover, in light of the high frequency of NTM infections detected in lesioned cattle of sub-Saharan Africa, other *Mycobacterium* spp. might pose a considerable health threat to humans and contaminated beef can be efficiently removed through abattoir meat inspections.

Although meat inspection is theoretically enforced in many countries, it is often not correctly accomplished and poorly implemented in rural areas. Losses from meat confiscations can generally not be refunded by the government. This situation promotes fraud and (illegal) private slaughter of animals. To overcome this problem, community funds could be established and financed by meat traders whose animals are subjected to meat inspection. Compensations for cattle condemnations could then be disbursed from these funds.

Interventions promoting milk pasteurization and meat inspection should specifically target remote areas where the population appears to be most frequently affected by zoonotic diseases [26]. However, it may be particularly difficult to eliminate the problem of private slaughter in rural settings. Also, implementation of pasteurization schemes may not be a trivial task as wood is often scarce and a precious resource, especially in arid zones. To overcome this problem it may be possible to concentrate milk pasteurization at specific points. The establishment of a dairy cooperative in the peri-urban region of Bamako where milk from small scale producers was pooled and pasteurized proved to cost-effectively promote the production of healthy dairy products (www.laitsain.com; [280]). However, in rural areas and pastoralist communities, close contact between humans and animals may pose another important risk factor for zoonotic transmission of *M. bovis*. In this respect, the promotion of public awareness would be most crucial to prevent the zoonotic transmission of *M. bovis*.

Bovine tuberculosis control in cattle

Implementation of control schemes for BTB in African cattle is a major challenge, starting with the difficulty of disease surveillance in livestock. A major problem is the lack of adequate diagnostic tests. Despite a number of drawbacks, SICCT is still the most frequently used test for ante-mortem diagnosis of BTB in Africa and to date, no satisfactory serological tests exist, which would allow the detection of *M. bovis* infected but SICCT anergic animals. Our finding of the better performance of SICCT in Chad and probably many other countries of sub-Saharan Africa if the cut-off for positive test interpretation is lowered to > 2 mm is therefore of high practical relevance [59,60]. A lowered cut-off will increase the test sensitivity, albeit at the cost of a decreased specificity. However, in view of the economic consequences of BTB in Africa we believe that the cost of a false-negative test outcome exceeds the cost of a false positive outcome by several folds [59,60]. Therefore, a high sensitivity of a diagnostic test to detect more of the infected cattle may be more important than a high specificity.

As mentioned previously, post mortem examination at abattoirs are primarily of importance from a food-safety perspective. Identification of the origin of animals exhibiting gross visible lesions is currently not possible in Africa, due to poor documentation. However, tracing of herd origin could be achieved e.g., by earmarking animals in the herds, by the establishment of sale or transport certificates and by the construction of databases of known herds and their location. It may be particularly difficult to get reliable information on the latter for moving populations although transhumant herds usually migrate along reasonably well defined routes [222]. Such a system would allow a detailed nation-wide surveillance of BTB and relatively rapid and targeted actions in regions and herds with high disease prevalence. It could thus effectively complement ante-mortem surveillance programs in cattle herds.

It has been shown that joint human- and animal health interventions in remote areas of Africa were significantly more successful than independent delivery of health services; this in terms of both, the total amount of individuals who received treatment as well as the cost-effectiveness of the intervention programs [281].

Similarly, delivery of public health services, campaigns to raise awareness for the risk factors of zoonotic disease transmission, livestock vaccination programs, cattle marking, the construction of herd specific databases and possibly ante-mortem testing of animals in herds could be conducted in a combined manner. To further reduce costs, diagnostic testing of cattle could target specific risk groups such as old cows. Importantly, such health interventions should also aim at improving the relationship between governmental institutions and the pastoralist communities as they are so far to a large extend excluded from many social services. Still, their contribution to the national economy is extensive and may be considerably increasing in the future, especially in view of the rapid population growth in Africa and the resulting larger demand for dairy products and meat. Finally, extensive livestock production systems also represent a method to fight the problem of BTB in Africa as they generally show a lower prevalence of the disease compared to intensive production systems. This also includes the keeping of local or cross breeds, which may be less susceptible to BTB than exotic breeds.

As outlined before, establishment of laboratories for regular molecular typing of MTBC strains would be of great value for BTB surveillance in Africa. Connecting molecular epidemiology and herd tracing as it was explained above could further enhance the power of molecular epidemiological investigations. The nation-wide population structure and distribution of *M. bovis* strains could be resolved. That is, the potential association of specific geographical regions within a country and bacterial genotypes could be investigated. Our molecular typing results on the Af1 clonal complex revealed the geographical localization on a country level of strains of *M. bovis* in West Africa; however, the same may be true on a more regional scale. Molecular epidemiology could also allow the detection of temporal changes within the population structure of *M. bovis*; this includes e.g. movements of bacterial clones from one region to another or changes within the population structure during interventions campaigns. Most importantly however, regular molecular typing of strains of *M. bovis* isolated from cattle carcasses at abattoirs and the possibility to trace the origin of the animals would allow the early detection of disease outbreaks, their localization and the identification of the specific strains involved.

The geographical localization of strains of *M. bovis* within the Af1 clonal complex [60] suggests that BTB control programs in their initial phase can be implemented on a national level without consideration of neighboring countries. However, the large spread of distinct clonal complexes of *M. bovis* in Africa highlights that cross border movements of cattle could become important at later stages of BTB control campaigns when disease prevalence is decreasing. In this respect, the establishment of concerted disease control schemes for all countries that are affected by the same clonal groups of *M. bovis* may be most effective. The recently founded BTB network in Africa could play an important role to coordinate such interventions. It can also help to raise awareness for the problem of BTB in governmental institutions and help to efficiently share the available resources for BTB control.

Molecular typing results indicated that several European countries may have actually exported BTB to Africa, presumably during the colonial period. This raises the question to what extend these countries should contribute in helping to fight the disease in Africa. This is especially important in view of the financial obstacles that hinder the implementation of control programs. But solid evidence for a European source of *M. bovis* in Africa requires further genomic analysis.

Control in wildlife

Our knowledge about the importance of *M. bovis* infections in wildlife in Africa is limited. The disease could have considerable consequences for the wildlife populations and the entire ecosystem. Moreover, it could pose an important threat to disease eradication programs in livestock as it is the case in the UK, Ireland and New Zealand [39,95,96]

In Africa, contact between wildlife and livestock is difficult to control. In sedentary livestock production systems, it could be prevented using e.g.

perimeter fences but occasional direct contacts or contact between infectious wildlife excreta and cattle is almost unavoidable in extensive transhumant livestock production systems. Direct interventions in wildlife populations e.g. through test and slaughter schemes seem to be difficult to accomplish [20]. Moreover, assuming a relatively low transmission rate between wildlife and livestock, the intervention expenses may be exceedingly high and disproportional to the revenues for the livestock sector. In such a scenario, vaccination of wildlife and/or livestock may be the most cost-effective intervention. However, administration of vaccines to wildlife populations might be very difficult and as mentioned previously, the currently available vaccines are not satisfactorily efficient [282]. Moreover, vaccinated and infected cattle cannot be distinguished using the presently available diagnostic tests. Research into the development of new vaccines and diagnostic tests therefore needs to be highly coordinated in order to allow a differential diagnosis for uninfected, infected and vaccinated individuals [155,282]. A major strategy for vaccine development is the improvement of the protective efficacy of BCG by boosting it with a second subunit vaccine, consisting of a cocktail of multiple antigens or a fusion protein [282,283]. It is improbable that a similar strategy would be usable for the vaccination of cattle or wildlife in Africa, considering the difficulties to retrieve a specific cattle herd, let alone an individual wild animal. A more suitable approach for domestic cattle in Africa may be neonatal vaccination. BCG vaccination of 1-day-old calves was shown to induce significant protection against *M. bovis* challenge in several studies [7,178]. Neonatal vaccination may be especially useful in African settings as it can circumvent potential problems associated with the exposure to environmental NTM [7].

Recommendations for future research

In many countries and specifically rural areas of Africa, still not much is known about the prevalence of BTB. Baseline studies, assessing the SICCT reactor prevalence in the herds would be highly needed. Such surveys should ideally be conducted in a coordinated manner, investigating both, animal reactor prevalence and human exposure to tuberculosis.

Current research into the development of new diagnostic tests, mainly aims at increasing test sensitivities for the detection of animals at very early infection stages. However, considering the characteristics of BTB in Africa and other developing countries and in order to most effectively target the major global burden of BTB, future research should also aim at developing serological tests, which allow the identification of anergic animals.

Most of the molecular epidemiological investigations in Africa so far only included few slaughterhouses in a country. New studies should survey several locations; moreover, tracing of the herd of origin would allow investigating the distribution of *M. bovis* strains inside a country. This approach would be most useful if implemented within a regular slaughterhouse surveillance scheme.

In addition, it would be particularly interesting to specifically investigate the strains and prevalence of BTB in cattle submitted to private slaughter compared to cattle sent to abattoirs.

To elucidate the global phylogeny of *M bovis* it would be necessary to sequence the genomes of different, distantly related strains of *M. bovis*. Our method of identifying clonal complexes offers an approach to recognize distinct clades throughout the world; representatives of each clade could then be selected for genome sequencing. Comparison of the sequence data would then allow the identification of SNPs, which could be used to reconstruct the phylogeny and to clarify the origin of *M. bovis*.

From our work and other studies, the importance of NTM infections in African cattle has become evident. It still remains to be proven that NTM can be a primary lesion causing agent. Artificial infections in animal models or in cattle could help to answer this question. Subsequently, the possibility of transmission of NTM strains should be explored and specific molecular epidemiological techniques may have to be invented.

Much more needs to be known about the influence of cattle husbandry and the livestock production strategies of nomadic pastoralists. Such studies could help to identify routes and risk factors for the transmission of *M. bovis* between animals and between animals and humans.

Economic investigations on BTB in Africa including cost and revenue analyses for all sectors are a prerequisite for the elaboration of intervention and control strategies. Almost no investigations have been conducted in this respect. Moreover, considering the high amount of NTM strains isolated from cattle of sub-Saharan Africa it may be worthwhile to assess the economic relevance of NTM infections in Africa.

Screening and testing of wildlife populations would help to assess the prevalence of *M. bovis* infections in wildlife and to estimate the transmission rate between cattle and wildlife. This in turn would allow to build up *M. bovis* transmission models that incorporate the wildlife reservoir and allow economic considerations concerning interventions in wildlife populations.

There is evidence that zoonotic transmission of *M. bovis* is mainly observed in pastoral areas. Future research should therefore specifically target this risk group in order to assess the importance of zoonotic *M. bovis* transmissions.

References

1. Smith NH, Gordon SV, de la Rua-Domenech R, Clifton-Hadley RS, Hewinson RG: **Bottlenecks and broomsticks: the molecular evolution of Mycobacterium bovis.** *Nature Reviews Microbiology* 2006, **4:** 670-681.

2. Wirth T, Hildebrand F, Allix-Beguec C, Wolbeling F, Kubica T, Kremer K *et al.*: **Origin, spread and demography of the Mycobacterium tuberculosis complex.** *PLoS Pathog* 2008, **4:** e1000160.

3. Hershberg R, Lipatov M, Small PM, Sheffer H, Niemann S, Homolka S *et al.*: **High functional diversity in Mycobacterium tuberculosis driven by genetic drift and human demography.** *PLoS Biol* 2008, **6:** e311.

4. Young DB, Perkins MD, Duncan K, Barry CE: **Confronting the scientific obstacles to global control of tuberculosis.** *Journal of Clinical Investigation* 2008, **118:** 1255-1265.

5. Ayele WY, Neill SD, Zinsstag J, Weiss MG, Pavlik I: **Bovine tuberculosis: an old disease but a new threat to Africa.** *Int J Tuberc Lung Dis* 2004, **8:** 924-937.

6. Cassidy JP: **The pathogenesis and pathology of bovine tuberculosis with insights from studies of tuberculosis in humans and laboratory animal models.** *Vet Microbiol* 2006, **112:** 151-161.

7. Hope JC, Villarreal-Ramos B: **Bovine TB and the development of new vaccines.** *Comp Immunol Microbiol Infect Dis* 2008, **31:** 77-100.

8. Coetzer JAW, Tustin RC: *Infectious diseases of livestock*, 2nd ed edn. Oxford: Oxford University Press; 2004.

9. Kubica T, Agzamova R, Wright A, Rakishev G, Rusch-Gerdes S, Niemann S: **Mycobacterium bovis isolates with M. tuberculosis specific characteristics.** *Emerg Infect Dis* 2006, **12:** 763-765.

10. Glickman MS, Jacobs WR, Jr.: **Microbial pathogenesis of Mycobacterium tuberculosis: dawn of a discipline.** *Cell* 2001, **104:** 477-485.

11. Steingart KR, Ng V, Henry M, Hopewell PC, Ramsay A, Cunningham J *et al.*: **Sputum processing methods to improve the sensitivity of smear microscopy for tuberculosis: a systematic review.** *Lancet Infect Dis* 2006, **6:** 664-674.

12. Cole ST: **Comparative and functional genomics of the Mycobacterium tuberculosis complex.** *Microbiology* 2002, **148:** 2919-2928.

13. Niemann S, Richter E, Rusch-Gerdes S: **Differentiation among members of the Mycobacterium tuberculosis complex by molecular and biochemical features: evidence for two pyrazinamide-susceptible subtypes of M. bovis.** *J Clin Microbiol* 2000, **38:** 152-157.

14. Huard RC, Fabre M, de HP, Lazzarini LC, van SD, Cousins D *et al.*: **Novel genetic polymorphisms that further delineate the phylogeny of the Mycobacterium tuberculosis complex.** *J Bacteriol* 2006, **188:** 4271-4287.

15. Boddinghaus B, Rogall T, Flohr T, Blocker H, Bottger EC: **Detection and identification of mycobacteria by amplification of rRNA.** *J Clin Microbiol* 1990, **28:** 1751-1759.

16. Durr PA, Hewinson RG, Clifton-Hadley RS: **Molecular epidemiology of bovine tuberculosis - I. Mycobacterium bovis genotyping.** *Revue Scientifique et Technique de l Office International des Epizooties* 2000, **19:** 675-688.

17. Mostrom P, Gordon M, Sola C, Ridell M, Rastogi N: **Methods used in the molecular epidemiology of tuberculosis.** *Clin Microbiol Infect* 2002, **8:** 694-704.

18. Pinsky BA, Banaei N: **Multiplex real-time PCR assay for rapid identification of Mycobacterium tuberculosis complex members to the species level.** *Journal of Clinical Microbiology* 2008, **46:** 2241-2246.

19. Zinsstag J, Schelling E, Roth F, Kazwala RR: **Economics of bovine tuberculosis.** In *Mycobacterium bovis Infection in Animals and Humans.* 2 edition. Edited by Thoen CO, Steele JH, Gilsdorf MJ. Blackwell Publishing; 2006.

20. Renwick AR, White PC, Bengis RG: **Bovine tuberculosis in southern African wildlife: a multi-species host-pathogen system.** *Epidemiol Infect* 2006, 1-12.

21. Michel AL, Bengis RG, Keet DF, Hofmeyr M, Klerk LM, Cross PC *et al.*: **Wildlife tuberculosis in South African conservation areas: implications and challenges.** *Vet Microbiol* 2006, **112:** 91-100.

22. Bengis RG, Kock RA, Fischer J: **Infectious animal diseases: the wildlife/livestock interface.** *Rev Sci Tech* 2002, **21:** 53-65.

23. Thoen C, Lobue P, de K, I: **The importance of Mycobacterium bovis as a zoonosis.** *Vet Microbiol* 2006, **112:** 339-345.

24. **Zoonotic tuberculosis (Mycobacterium bovis): memorandum from a WHO meeting (with the participation of FAO).** *Bull World Health Organ* 1994, **72:** 851-857.

25. Rodwell TC, Moore M, Moser KS, Brodine SK, Strathdee SA: **Tuberculosis from Mycobacterium bovis in binational communities, United States.** *Emerg Infect Dis* 2008, **14:** 909-916.

26. WHO. The Control of Neglected Zoonotic Diseases. A route to poverty allevation. 2006. Geneva, World Health Organization.
Ref Type: Report

27. Lobue P: **Public Health Significance of *M. bovis*.** In *Mycobacterium bovis infection in animals and humans.* 2nd edition. Edited by Thoen CO, Steele JH, Gilsdorf MJ. Ames, Iowa 50014, USA: Blackwell Publishing; 2006:6-12.

28. Otte MJ, Chilonda P: **Cattle and small ruminant production systems in sub-Saharan Africa - A systematic review.** *Food and Agriculture Organization of the United Nations, Rome 2002* 2002.

29. Thornton PK, Kruska RL, Henninger N, Kristjanson PM, Reid RS, Atieno F *et al.*: *Mapping Poverty and Livestock in the Developing World.* Nairobi, Kenya: International Livestock Research Institute (ILRI); 2002.

30. Kaneene JB, Pfeiffer D: **Epidemiology of *Mycobacterium bovis*.** In *Mycobacterium bovis infection in animals and humans.* 2 edition. Edited by Thoen CO, Steele JH, Gilsdorf MJ. Ames, Iowa 50014, USA: Blackwell Publishing; 2006:34-48.

31. Ameni G, Aseffa A, Engers H, Young D, Hewinson G, Vordermeier M: **Cattle husbandry in Ethiopia is a predominant factor affecting the pathology of bovine tuberculosis and gamma interferon responses to mycobacterial antigens.** *Clin Vaccine Immunol* 2006, **13:** 1030-1036.

32. Cosivi O, Grange JM, Daborn CJ, Raviglione MC, Fujikura T, Cousins D *et al.*: **Zoonotic tuberculosis due to Mycobacterium bovis in developing countries.** *Emerg Infect Dis* 1998, **4:** 59-70.

33. Cleaveland S, Shaw DJ, Mfinanga SG, Shirima G, Kazwala RR, Eblate E *et al.*: **Mycobacterium bovis in rural Tanzania: risk factors for infection in human and cattle populations.** *Tuberculosis (Edinb)* 2007, **87:** 30-43.

34. Kazwala RR, Kambarage DM, Daborn CJ, Nyange J, Jiwa SF, Sharp JM: **Risk factors associated with the occurrence of bovine tuberculosis in cattle in the Southern Highlands of Tanzania.** *Vet Res Commun* 2001, **25:** 609-614.

35. Munyeme M, Muma JB, Samui KL, Skjerve E, Nambota AM, Phiri IG *et al.*: **Prevalence of bovine tuberculosis and animal level risk factors for indigenous cattle under different grazing strategies in the livestock/wildlife interface areas of Zambia.** *Trop Anim Health Prod* 2008.

36. Oloya J, Opuda-Asibo J, Djonne B, Muma JB, Matope G, Kazwala R *et al.*: **Responses to tuberculin among Zebu cattle in the transhumance regions of Karamoja and Nakasongola district of Uganda.** *Trop Anim Health Prod* 2006, **38:** 275-283.

37. Cheeseman CL, Wilesmith JW, Stuart FA: **Tuberculosis: the disease and its epidemiology in the badger, a review.** *Epidemiol Infect* 1989, **103:** 113-125.

38. O'Brien DJ, Schmitt SM, Fitzgerald SD, Berry DE, Hickling GJ: **Managing the wildlife reservoir of Mycobacterium bovis: the Michigan, USA, experience.** *Vet Microbiol* 2006, **112:** 313-323.

39. Coleman JD, Cooke MM: **Mycobacterium bovis infection in wildlife in New Zealand.** *Tuberculosis (Edinb)* 2001, **81:** 191-202.

40. Porphyre T, Stevenson MA, McKenzie J: **Risk factors for bovine tuberculosis in New Zealand cattle farms and their relationship with possum control strategies.** *Prev Vet Med* 2008, **86:** 93-106.

41. Ameni G, Aseffa A, Engers H, Young D, Gordon S, Hewinson G *et al.*: **Both prevalence and severity of pathology of bovine tuberculosis are higher in Holsteins than in zebu breeds under field cattle husbandry in central Ethiopia.** *Clin Vaccine Immunol* 2007.

42. Ameni G, Aseffa A, Engers H, Young D, Gordon S, Hewinson G *et al.*: **High prevalence and increased severity of pathology of bovine tuberculosis in Holsteins compared to zebu breeds under field cattle husbandry in central Ethiopia.** *Clin Vaccine Immunol* 2007, **14:** 1356-1361.

43. Diguimbaye-Djaïbe C, Hilty M, Ngandolo R, Mahamat HH, Pfyffer GE, Baggi F *et al.*: **Mycobacterium bovis isolates from tuberculous lesions in Chadian zebu carcasses.** *Emerg Infect Dis* 2006, **12:** 769-771.

44. Doutre MP: **[Note concerning the recent cases of bovine tuberculosis (Myobacterium bovis) at the abattoir of Dakar].** *Rev Elev Med Vet Pays Trop* 1976, **29:** 309-311.

45. Zinsstag J, Kazwala RR, Cadmus S, Ayanwale L: ***Mycobacterium bovis* in Africa.** In *Mycobacterium bovis Infection in Animals and Humans.* 2 edition. Edited by Thoen CO, Steele JH, Gilsdorf MJ. Blackwell Publishing; 2006.

46. Sidibé SS, Dicko NA, Fané A, Doumbia RM, Sidibé CK, Kanté S *et al.*: **Tuberculose bovine au Mali : résultats d'une enquête épidemiologique dans les élevages laitiers de la zone périurbaine du district de Bamako.** *Revue Élev Méd vét Pays trop* 2003, **56:** 115-120.

47. Müller B, Steiner B, Bonfoh B, Fane A, Smith NH, Zinsstag J: **Molecular characterisation of Mycobacterium bovis isolated from**

cattle slaughtered at the Bamako abattoir in Mali. *BMC Vet Res* 2008, **4:** 26.

48. Delafosse A, Traore A, Kone B: **[Isolation of pathogenic Mycobacterium strains in cattle slaughtered in the abattoir of Bobo-Dioulasso, Burkina Faso].** *Rev Elev Med Vet Pays Trop* 1995, **48:** 301-306.

49. Vekemans M, Cartoux M, Diagbouga S, Dembele M, Kone B, Delafosse A *et al.*: **Potential source of human exposure to Mycobacterium bovis in Burkina Faso, in the context of the HIV epidemic.** *Clin Microbiol Infect* 1999, **5:** 617-621.

50. Rey JL, Pestiaux JL, Bichat B, Meyer L: **[Bacteriological study of 174 mycobacterial strains isolated from tuberculosis patients in Niger].** *Ann Soc Belg Med Trop* 1982, **62:** 55-60.

51. Addo K, Owusu-Darko K, Yeboah-Manu D, Caulley P, Minamikawa M, Bonsu F *et al.*: **Mycobacterial species causing pulmonary tuberculosis at the korle bu teaching hospital, accra, ghana.** *Ghana Med J* 2007, **41:** 52-57.

52. Cadmus S, Palmer S, Okker M, Dale J, Gover K, Smith N *et al.*: **Molecular analysis of human and bovine tubercle bacilli from a local setting in Nigeria.** *J Clin Microbiol* 2006, **44:** 29-34.

53. Kamerbeek J, Schouls L, Kolk A, van Agterveld M, van Soolingen D, Kuijper S *et al.*: **Simultaneous detection and strain differentiation of Mycobacterium tuberculosis for diagnosis and epidemiology.** *J Clin Microbiol* 1997, **35:** 907-914.

54. Idigbe EO, Anyiwo CE, Onwujekwe DI: **Human pulmonary infections with bovine and atypical mycobacteria in Lagos, Nigeria.** *J Trop Med Hyg* 1986, **89:** 143-148.

55. Mawak J, Gomwalk N, Bello C, Kandakai-Olukemi Y: **Human pulmonary infections with bovine and environment (atypical) mycobacteria in jos, Nigeria.** *Ghana Med J* 2006, **40:** 132-136.

56. Schelling E, Diguimbaye C, Daoud S, Daugla DM, Bidjeh K, Tanner M *et al.*: **La tuberculose causée par mycobacterium bovis: résultats préliminaires obtenus chez les pasteurs nomades foulbés et arabes dans le chari-baguirmi au Tchad.** *Sermpervira* 2000, **8:** 44-55.

57. Delafosse A, Goutard F, Thébaud E: **Epidémiologie de la tuberculose et de la brucellose des bovins en zone périurbaine d'Abéché, Tchad [Epidemiology of bovine tuberculosis and brucellosis on the Periphery of Abeche, Chad].** *Rev élev méd vét pays trop* 2002, **55:** 5-13.

58. Diguimbaye-Djaïbe C, Vincent V, Schelling E, Hilty M, Ngandolo R, Mahamat HH *et al.*: **Species identification of non-tuberculous**

mycobacteria from humans and cattle of Chad. *Schweiz Arch Tierheilkd* 2006, **148:** 251-256.

59. Ngandolo BNR, Müller B, Diguimbaye-Djaïbe C, Schiller I, Marg-Haufe B, Cagiola M *et al.*: **Comparative assessment of fluorescence polarization and tuberculin skin testing for the diagnosis of bovine tuberculosis in Chadian cattle.** *Prev Vet Med* 2009, **89:** 81-89.

60. Müller B, Vounatsou P, Ngandolo BN, Diguimbaye-Djaïbe C, Schiller I, Marg-Haufe B *et al.*: **Bayesian receiver operating characteristic estimation of multiple tests for diagnosis of bovine tuberculosis in Chadian cattle.** *PLoS ONE* 2009, **4:** e8215.

61. Njanpop-Lafourcade BM, Inwald J, Ostyn A, Durand B, Hughes S, Thorel MF *et al.*: **Molecular typing of Mycobacterium bovis isolates from Cameroon.** *J Clin Microbiol* 2001, **39:** 222-227.

62. Sulieman MS, Hamid ME: **Identification of acid fast bacteria from caseous lesions in cattle in Sudan.** *J Vet Med B Infect Dis Vet Public Health* 2002, **49:** 415-418.

63. Asseged B, Woldesenbet Z, Yimer E, Lemma E: **Evaluation of abattoir inspection for the diagnosis of Mycobacterium bovis infection in cattle at Addis Ababa abattoir.** *Trop Anim Health Prod* 2004, **36:** 537-546.

64. Teklul A, Asseged B, Yimer E, Gebeyehu M, Woldesenbet Z: **Tuberculous lesions not detected by routine abattoir inspection: the experience of the Hossana municipal abattoir, southern Ethiopia.** *Rev Sci Tech* 2004, **23:** 957-964.

65. Ameni G, Hewinson G, Aseffa A, Young D, Vordermeier M: **Appraisal of interpretation criteria for the comparative intradermal tuberculin test for the diagnosis of bovine tuberculosis in central Ethiopia.** *Clin Vaccine Immunol* 2008.

66. Ameni G, Aseffa A, Sirak A, Engers H, Young DB, Hewinson RG *et al.*: **Effect of skin testing and segregation on the prevalence of bovine tuberculosis, and molecular typing of Mycobacterium bovis, in Ethiopia.** *Vet Rec* 2007, **161:** 782-786.

67. Berg S, Firdessa R, Habtamu M, Gadisa E, Mengistu A, Yamuah L *et al.*: **The burden of mycobacterial disease in ethiopian cattle: implications for public health.** *PLoS ONE* 2009, **4:** e5068.

68. Oloya J, Muma JB, Opuda-Asibo J, Djonne B, Kazwala R, Skjerve E: **Risk factors for herd-level bovine-tuberculosis seropositivity in transhumant cattle in Uganda.** *Prev Vet Med* 2007, **80:** 318-329.

69. Oloya J, Kazwala R, Lund A, Opuda-Asibo J, Demelash B, Skjerve E *et al.*: **Characterisation of mycobacteria isolated from slaughter cattle in pastoral regions of Uganda.** *BMC Microbiol* 2007, **7:** 95.

70. Asiimwe BB, Asiimwe J, Kallenius G, Ashaba FK, Ghebremichael S, Joloba M *et al.*: **Molecular characterisation of Mycobacterium bovis isolates from cattle carcases at a city slaughterhouse in Uganda.** *Vet Rec* 2009, **164:** 655-658.

71. Mposhy M, Binemomadi C, Mudakikwa B: **Incidence of Bovine Tuberculosis on the Health of the Populations of North-Kivu (Zaire).** *Revue D Elevage et de Medecine Veterinaire des Pays Tropicaux* 1983, **36:** 15-18.

72. Kang'ethe EK, Ekuttan CE, Kimani VN: **Investigation of the prevalence of bovine tuberculosis and risk factors for human infection with bovine tuberculosis among dairy and non-dairy farming neighbour households in Dagoretti Division, Nairobi, Kenya.** *East Afr Med J* 2007, **84:** S92-S95.

73. Sapolsky RM, Else JG: **Bovine tuberculosis in a wild baboon population: epidemiological aspects.** *J Med Primatol* 1987, **16:** 229-235.

74. Jiwa SFH, Kazwala RR, Aboud AAO, Kalaye WJ: **Bovine tuberculosis in the Lake Victoria Zone of Tanzania and its possible consequences for human health in the HIV/AIDS era.** *Veterinary Research Communications* 1997, **21:** 533-539.

75. Weinhaupl I, Schopf KC, Khaschabi D, Kapaga AM, Msami HM: **Investigations on the prevalence of bovine tuberculosis and brucellosis in dairy cattle in Dar es Salaam region and in zebu cattle in Lugoba area, Tanzania.** *Tropical Animal Health and Production* 2000, **32:** 147-154.

76. Shirima GM, Kazwala RR, Kambarage DM: **Prevalence of bovine tuberculosis in cattle in different farming systems in the eastern zone of Tanzania.** *Preventive Veterinary Medicine* 2003, **57:** 167-172.

77. Kazwala RR, Daborn CJ, Kusiluka LJM, Jiwa SFH, Sharp JM, Kambarage DM: **Isolation of Mycobacterium species from raw milk of pastoral cattle of the Southern Highlands of Tanzania.** *Tropical Animal Health and Production* 1998, **30:** 233-239.

78. Kazwala RR, Kusiluka LJ, Sinclair K, Sharp JM, Daborn CJ: **The molecular epidemiology of Mycobacterium bovis infections in Tanzania.** *Vet Microbiol* 2005.

79. Cleaveland S, Mlengeya T, Kazwala RR, Michel A, Kaare MT, Jones SL *et al.*: **Tuberculosis in Tanzanian wildlife.** *J Wildl Dis* 2005, **41:** 446-453.

80. Bedard BG, Martin SW, Chinombo D: **A Prevalence Study of Bovine Tuberculosis and Brucellosis in Malawi.** *Preventive Veterinary Medicine* 1993, **16:** 193-205.

81. Cook AJC, Tuchili LM, Buve A, Foster SD, GodfreyFaussett P, Pandey GS *et al.*: **Human and bovine tuberculosis in the Monze District of Zambia - A cross-sectional study.** *British Veterinary Journal* 1996, **152:** 37-46.

82. Phiri AM: **Common conditions leading to cattle carcass and offal condemnations at 3 abattoirs in the Western Province of Zambia and their zoonotic implications to consumers.** *Journal of the South African Veterinary Association-Tydskrif Van Die Suid-Afrikaanse Veterinere Vereniging* 2006, **77:** 28-32.

83. Munyeme M, Muma JB, Skjerve E, Nambota AM, Phiri IGK, Samui KL *et al.*: **Risk factors associated with bovine tuberculosis in traditional cattle of the livestock/wildlife interface areas in the Kafue basin of Zambia.** *Preventive Veterinary Medicine* 2008, **85:** 317-328.

84. Zieger U, Pandey GS, Kriek NPJ, Cauldwell AE: **Tuberculosis in Kafue lechwe (Kobus leche kafuensis) and in a bushbuck (Tragelaphus scriptus) on a game ranch in Central Province, Zambia.** *Journal of the South African Veterinary Association-Tydskrif Van Die Suid-Afrikaanse Veterinere Vereniging* 1998, **69:** 98-101.

85. Munyeme M, Rigouts L, Shamputa IC, Muma JB, Tryland M, Skjerve E *et al.*: **Isolation and characterization of Mycobacterium bovis strains from indigenous Zambian cattle using Spacer oligonucleotide typing technique.** *BMC Microbiol* 2009, **9:** 144.

86. Michel AL, Hlokwe TM, Coetzee ML, Mare L, Connoway L, Rutten VP *et al.*: **High Mycobacterium bovis genetic diversity in a low prevalence setting.** *Vet Microbiol* 2008, **126:** 151-159.

87. Quirin R, Rasolofo V, Andriambololona R, Ramboasolo A, Rasolonavalona T, Raharisolo C *et al.*: **Validity of intradermal tuberculin testing for the screening of bovine tuberculosis in Madagascar.** *Onderstepoort Journal of Veterinary Research* 2001, **68:** 231-238.

88. Rasolofo R, V, Quirin R, Rapaoliarijaona A, Rakotoaritahina H, Vololonirina EJ, Rasolonavalona T *et al.*: **Usefulness of restriction fragment length polymorphism and spoligotyping for epidemiological studies of Mycobacterium bovis in Madagascar: description of new genotypes.** *Vet Microbiol* 2006, **114:** 115-122.

89. Gilsdorf MJ, Ebel ED, Disney TW: **Benefit and Cost Assessment of the U.S. Bovine Tuberculosis Eradication program.** In *Mycobacterium bovis infection in animals and humans.* 2 edition. Edited by Thoen CO, Steele JH, Gilsdorf MJ. Ames, Iowa 50014, USA: Blackwell Publishing; 2006:89-99.

90. Bernues A, Manrique E, Maza MT: **Economic evaluation of bovine brucellosis and tuberculosis eradication programmes in a mountain area of Spain.** *Prev Vet Med* 1997, **30:** 137-149.

91. Munag'andu HM, Siamudaala VM, Nambota A, Bwalya JM, Munyeme M, Mweene AS *et al.*: **Disease constraints for utilization of the African buffalo (Syncerus caffer) on game ranches in Zambia.** *Jpn J Vet Res* 2006, **54:** 3-13.

92. Steinfeld H, Wassenaar T, Jutzi S: **Livestock production systems in developing countries: status, drivers, trends.** *Rev Sci Tech* 2006, **25:** 505-516.

93. Pollock JM, Rodgers JD, Welsh MD, McNair J: **Pathogenesis of bovine tuberculosis: the role of experimental models of infection.** *Vet Microbiol* 2006, **112:** 141-150.

94. Roth F, Zinsstag J, Orkhon D, Chimed-Ochir G, Hutton G, Cosivi O *et al.*: **Human health benefits from livestock vaccination for brucellosis: case study.** *Bull World Health Organ* 2003, **81:** 867-876.

95. Abernethy DA, Denny GO, Menzies FD, McGuckian P, Honhold N, Roberts AR: **The Northern Ireland programme for the control and eradication of Mycobacterium bovis.** *Vet Microbiol* 2006, **112:** 231-237.

96. Delahay RJ, De Leeuw AN, Barlow AM, Clifton-Hadley RS, Cheeseman CL: **The status of Mycobacterium bovis infection in UK wild mammals: a review.** *Vet J* 2002, **164:** 90-105.

97. Serraino A, Marchetti G, Sanguinetti V, Rossi MC, Zanoni RC, Catozzi L *et al.*: **Monitoring of transmission of tuberculosis between wild boars and cattle: Genotypical analysis of strains by molecular epidemiology techniques.** *Journal of Clinical Microbiology* 1999, **37:** 2766-2771.

98. Aranaz A, de JL, Montero N, Sanchez C, Galka M, Delso C *et al.*: **Bovine tuberculosis (Mycobacterium bovis) in wildlife in Spain.** *J Clin Microbiol* 2004, **42:** 2602-2608.

99. Woodford MH: **Tuberculosis in wildlife in the Ruwenzori National Park Uganda (part I).** *Trop Anim Health Prod* 1982, **14:** 81-88.

100. Woodford MH: **Tuberculosis in wildlife in the Ruwenzori National Park, Uganda (Part II).** *Trop Anim Health Prod* 1982, **14:** 155-160.

101. de IR-D: **Human Mycobacterium bovis infection in the United Kingdom: Incidence, risks, control measures and review of the zoonotic aspects of bovine tuberculosis.** *Tuberculosis (Edinb)* 2006, **86:** 77-109.

102. Cousins DV, Williams SN, Dawson DJ: **Tuberculosis due to Mycobacterium bovis in the Australian population: DNA typing of isolates, 1970-1994.** *Int J Tuberc Lung Dis* 1999, **3:** 722-731.

103. Lobue PA, Betacourt W, Peter C, Moser KS: **Epidemiology of Mycobacterium bovis disease in San Diego County, 1994-2000.**

International Journal of Tuberculosis and Lung Disease 2003, **7:** 180-185.

104. Godreuil S, Torrea G, Terru D, Chevenet F, Diagbouga S, Supply P *et al.*: **First molecular epidemiology study of Mycobacterium tuberculosis in Burkina Faso.** *J Clin Microbiol* 2007, **45:** 921-927.

105. Eldholm V, Matee M, Mfinanga SG, Heun M, Dahle UR: **A first insight into the genetic diversity of Mycobacterium tuberculosis in Dar es Salaam, Tanzania, assessed by spoligotyping.** *BMC Microbiol* 2006, **6:** 76.

106. Shemko M, Yates M, Fang Z, Gibson A, Shetty N: **Molecular epidemiology of Mycobacterium tuberculosis in patients of Somalian and white ethnic origin attending an inner London clinic.** *Int J Tuberc Lung Dis* 2004, **8:** 186-193.

107. Bruchfeld J, Aderaye G, Palme IB, Bjorvatn B, Ghebremichael S, Hoffner S *et al.*: **Molecular epidemiology and drug resistance of Mycobacterium tuberculosis isolates from Ethiopian pulmonary tuberculosis patients with and without human immunodeficiency virus infection.** *J Clin Microbiol* 2002, **40:** 1636-1643.

108. Heyderman RS, Goyal M, Roberts P, Ushewokunze S, Zizhou S, Marshall BG *et al.*: **Pulmonary tuberculosis in Harare, Zimbabwe: analysis by spoligotyping.** *Thorax* 1998, **53:** 346-350.

109. Homolka S, Post E, Oberhauser B, George AG, Westman L, Dafae F *et al.*: **High genetic diversity among Mycobacterium tuberculosis complex strains from Sierra Leone.** *BMC Microbiol* 2008, **8:** 103.

110. Rigouts L, Maregeya B, Traore H, Collart JP, Fissette K, Portaels F: **Use of DNA restriction fragment typing in the differentiation of Mycobacterium tuberculosis complex isolates from animals and humans in Burundi.** *Tuber Lung Dis* 1996, **77:** 264-268.

111. Diguimbaye C, Hilty M, Ngandolo R, Mahamat HH, Pfyffer GE, Baggi F *et al.*: **Molecular characterization and drug resistance testing of Mycobacterium tuberculosis isolates from Chad.** *J Clin Microbiol* 2006, **44:** 1575-1577.

112. Mfinanga SG, Morkve O, Kazwala RR, Cleaveland S, Sharp MJ, Kunda J *et al.*: **Mycobacterial adenitis: role of Mycobacterium bovis, non-tuberculous mycobacteria, HIV infection, and risk factors in Arusha, Tanzania.** *East Afr Med J* 2004, **81:** 171-178.

113. Kazwala RR, Daborn CJ, Sharp JM, Kambarage DM, Jiwa SF, Mbembati NA: **Isolation of Mycobacterium bovis from human cases of cervical adenitis in Tanzania: a cause for concern?** *Int J Tuberc Lung Dis* 2001, **5:** 87-91.

114. Oloya J, Opuda-Asibo J, Kazwala R, Demelash AB, Skjerve E, Lund A *et al.*: **Mycobacteria causing human cervical lymphadenitis in**

pastoral communities in the Karamoja region of Uganda. *Epidemiol Infect* 2008, **136:** 636-643.

115. Rasolofo-Razanamparany V, Menard D, Rasolonavalona T, Ramarokoto H, Rakotomanana F, Auregan G *et al.*: **Prevalence of Mycobacterium bovis in human pulmonary and extra-pulmonary tuberculosis in Madagascar.** *Int J Tuberc Lung Dis* 1999, **3:** 632-634.

116. Kallenius G, Koivula T, Ghebremichael S, Hoffner SE, Norberg R, Svensson E *et al.*: **Evolution and clonal traits of Mycobacterium tuberculosis complex in Guinea-Bissau.** *J Clin Microbiol* 1999, **37:** 3872-3878.

117. Niobe-Eyangoh SN, Kuaban C, Sorlin P, Cunin P, Thonnon J, Sola C *et al.*: **Genetic biodiversity of Mycobacterium tuberculosis complex strains from patients with pulmonary tuberculosis in Cameroon.** *J Clin Microbiol* 2003, **41:** 2547-2553.

118. Asiimwe BB, Koivula T, Kallenius G, Huard RC, Ghebremichael S, Asiimwe J *et al.*: **Mycobacterium tuberculosis Uganda genotype is the predominant cause of TB in Kampala, Uganda.** *Int J Tuberc Lung Dis* 2008, **12:** 386-391.

119. Niemann S, Rusch-Gerdes S, Joloba ML, Whalen CC, Guwatudde D, Ellner JJ *et al.*: **Mycobacterium africanum subtype II is associated with two distinct genotypes and is a major cause of human tuberculosis in Kampala, Uganda.** *J Clin Microbiol* 2002, **40:** 3398-3405.

120. Sharaf-Eldin GS, Saeed NS, Hamid ME, Jordaan AM, van der Spuy GD, Warren RM *et al.*: **Molecular analysis of clinical isolates of Mycobacterium tuberculosis collected from patients with persistent disease in the Khartoum region of Sudan.** *J Infect* 2002, **44:** 244-251.

121. Koeck JL, Bernatas JJ, Gerome P, Fabre M, Houmed A, Herve V *et al.*: **[Epidemiology of resistance to antituberculosis drugs in Mycobacterium tuberculosis complex strains isolated from adenopathies in Djibouti. Prospective study carried out in 1999].** *Med Trop (Mars)* 2002, **62:** 70-72.

122. Cooksey RC, Abbadi SH, Woodley CL, Sikes D, Wasfy M, Crawford JT *et al.*: **Characterization of Mycobacterium tuberculosis complex isolates from the cerebrospinal fluid of meningitis patients at six fever hospitals in Egypt.** *J Clin Microbiol* 2002, **40:** 1651-1655.

123. Enarson DA: **Introduction.** In *Mycobacterium bovis infection in animals and humans.* 2 edition. Edited by Thoen CO, Steele JH, Gilsdorf MJ. Ames, Iowa 50014, USA: Blackwell Publishing; 2006:1-5.

124. Zinsstag J, Ould TM, Craig PS: **Editorial: health of nomadic pastoralists: new approaches towards equity effectiveness.** *Trop Med Int Health* 2006, **11:** 565-568.

125. Harris NB: **Molecular Techniques: Applications in Epidemiologic Studies.** In *Myctobacterium bovis infection in animals and humans.* 2 edition. Edited by Thoen CO, Steele JH, Gilsdorf MJ. Ames, Iowa 50014, USA: Blackwell Publishing; 2006:54-62.

126. van Soolingen D: **Molecular epidemiology of tuberculosis and other mycobacterial infections: main methodologies and achievements.** *Journal of Internal Medicine* 2001, **249:** 1-26.

127. Kanduma E, McHugh TD, Gillespie SH: **Molecular methods for Mycobacterium tuberculosis strain typing: a users guide.** *J Appl Microbiol* 2003, **94:** 781-791.

128. Haddad N, Masselot M, Durand B: **Molecular differentiation of Mycobacterium bovis isolates. Review of main techniques and applications.** *Res Vet Sci* 2004, **76:** 1-18.

129. Hermans PW, van SD, Bik EM, de Haas PE, Dale JW, van Embden JD: **Insertion element IS987 from Mycobacterium bovis BCG is located in a hot-spot integration region for insertion elements in Mycobacterium tuberculosis complex strains.** *Infect Immun* 1991, **59:** 2695-2705.

130. van der Zanden AG, Kremer K, Schouls LM, Caimi K, Cataldi A, Hulleman A *et al.*: **Improvement of differentiation and interpretability of spoligotyping for Mycobacterium tuberculosis complex isolates by introduction of new spacer oligonucleotides.** *J Clin Microbiol* 2002, **40:** 4628-4639.

131. Allix C, Walravens K, Saegerman C, Godfroid J, Supply P, Fauville-Dufaux M: **Evaluation of the epidemiological relevance of variable-number tandem-repeat genotyping of Mycobacterium bovis and comparison of the method with IS6110 restriction fragment length polymorphism analysis and spoligotyping.** *J Clin Microbiol* 2006, **44:** 1951-1962.

132. Aranaz A, Liebana E, Mateos A, Dominguez L, Vidal D, Domingo M *et al.*: **Spacer oligonucleotide typing of Mycobacterium bovis strains from cattle and other animals: A tool for studying epidemiology of tuberculosis.** *Journal of Clinical Microbiology* 1996, **34:** 2734-2740.

133. Clifton-Hadley RS, Inwald J, Hughes S, Palmer N, Sayers AR, Sweeney K *et al.*. Recent advances in DNA fingerprinting using spoligotyping - Epidemiological applications in bovine TB. 1998.
Ref Type: Generic

134. Costello E, O'Grady D, Flynn O, O'Brien R, Rogers M, Quigley F *et al.*: **Study of restriction fragment length polymorphism analysis and spoligotyping for epidemiological investigation of Mycobacterium bovis infection.** *J Clin Microbiol* 1999, **37:** 3217-3222.

135. Vitol I, Driscoll J, Kreiswirth B, Kurepina N, Bennett KP: **Identifying Mycobacterium tuberculosis complex strain families using spoligotypes.** *Infect Genet Evol* 2006, **6:** 491-504.

136. Brudey K, Driscoll JR, Rigouts L, Prodinger WM, Gori A, Al-Hajoj SA *et al.*: **Mycobacterium tuberculosis complex genetic diversity: mining the fourth international spoligotyping database (SpolDB4) for classification, population genetics and epidemiology.** *Bmc Microbiology* 2006, **6.**

137. Smith NH, Kremer K, Inwald J, Dale J, Driscoll JR, Gordon SV *et al.*: **Ecotypes of the Mycobacterium tuberculosis complex.** *J Theor Biol* 2005.

138. Frothingham R, Meeker-O'Connell WA: **Genetic diversity in the Mycobacterium tuberculosis complex based on variable numbers of tandem DNA repeats.** *Microbiology-Uk* 1998, **144:** 1189-1196.

139. Supply P, Mazars E, Lesjean S, Vincent V, Gicquel B, Locht C: **Variable human minisatellite-like regions in the Mycobacterium tuberculosis genome.** *Molecular Microbiology* 2000, **36:** 762-771.

140. Skuce RA, McCorry TP, McCarroll JF, Roring SMM, Scott AN, Brittain D *et al.*: **Discrimination of Mycobacterium tuberculosis complex bacteria using novel VNTR-PCR targets.** *Microbiology-Sgm* 2002, **148:** 519-528.

141. Filliol I, Ferdinand S, Negroni L, Sola C, Rastogi N: **Molecular typing of Mycobacterium tuberculosis based on variable number of tandem DNA repeats used alone and in association with spoligotyping.** *J Clin Microbiol* 2000, **38:** 2520-2524.

142. Hilty M, Diguimbaye C, Schelling E, Baggi F, Tanner M, Zinsstag J: **Evaluation of the discriminatory power of variable number tandem repeat (VNTR) typing of Mycobacterium bovis strains.** *Vet Microbiol* 2005, **109:** 217-222.

143. Kremer K, van Soolingen D, Frothingham R, Haas WH, Hermans PWM, Martin C *et al.*: **Comparison of methods based on different molecular epidemiological markers for typing of Mycobacterium tuberculosis complex strains: Interlaboratory study of discriminatory power and reproducibility.** *Journal of Clinical Microbiology* 1999, **37:** 2607-2618.

144. Gagneux S, Small PM: **Global phylogeography of Mycobacterium tuberculosis and implications for tuberculosis product development.** *Lancet Infect Dis* 2007, **7:** 328-337.

145. Brosch R, Gordon SV, Marmiesse M, Brodin P, Buchrieser C, Eiglmeier K *et al.*: **A new evolutionary scenario for the Mycobacterium tuberculosis complex.** *Proc Natl Acad Sci U S A* 2002, **99:** 3684-3689.

146. Kubica T, Rusch-Gerdes S, Niemann S: **Mycobacterium bovis subsp. caprae caused one-third of human M. bovis-associated tuberculosis cases reported in Germany between 1999 and 2001.** *J Clin Microbiol* 2003, **41:** 3070-3077.

147. Xavier EF, Seagar AL, Doig C, Rayner A, Claxton P, Laurenson I: **Human and animal infections with Mycobacterium microti, Scotland.** *Emerg Infect Dis* 2007, **13:** 1924-1927.

148. Cohan FM: **Bacterial species and speciation.** *Syst Biol* 2001, **50:** 513-524.

149. Mostowy S, Inwald J, Gordon S, Martin C, Warren R, Kremer K *et al.*: **Revisiting the evolution of Mycobacterium bovis.** *J Bacteriol* 2005, **187:** 6386-6395.

150. Gagneux S, Deriemer K, Van T, Kato-Maeda M, de Jong BC, Narayanan S *et al.*: **Variable host-pathogen compatibility in Mycobacterium tuberculosis.** *Proc Natl Acad Sci U S A* 2006.

151. Gutierrez MC, Brisse S, Brosch R, Fabre M, Omais B, Marmiesse M *et al.*: **Ancient origin and gene mosaicism of the progenitor of mycobacteriumtuberculosis.** *PLoS Pathog* 2005, **1:** e5.

152. Supply P, Warren RM, Banuls AL, Lesjean S, van der Spuy GD, Lewis LA *et al.*: **Linkage disequilibrium between minisatellite loci supports clonal evolution of Mycobacterium tuberculosis in a high tuberculosis incidence area.** *Mol Microbiol* 2003, **47:** 529-538.

153. Goodchild T, Clifton-Hadley R: **The Fall and Rise of Bovine Tuberculosis in Great Britain.** In *Mycobacterium bovis infection in animals and humans.* 2 edition. Edited by Thoen CO, Steele JH, Gilsdorf MJ. Ames, Iowa 50014, USA: Blackwell Publishing; 2006:100-116.

154. Smith NH, Dale J, Inwald J, Palmer S, Gordon SV, Hewinson RG *et al.*: **The population structure of Mycobacterium bovis in Great Britain: clonal expansion.** *Proc Natl Acad Sci U S A* 2003, **100:** 15271-15275.

155. Hewinson RG, Vordermeier HM, Smith NH, Gordon SV: **Recent advances in our knowledge of Mycobacterium bovis: a feeling for the organism.** *Vet Microbiol* 2006, **112:** 127-139.

156. Nguyen L, Pieters J: **The Trojan horse: survival tactics of pathogenic mycobacteria in macrophages.** *Trends Cell Biol* 2005, **15:** 269-276.

157. van der WN, Hava D, Houben D, Fluitsma D, van ZM, Pierson J *et al.*: **M. tuberculosis and M. leprae translocate from the phagolysosome to the cytosol in myeloid cells.** *Cell* 2007, **129:** 1287-1298.

158. Thoen CO, Barletta RG: **Pathogenesis of *Mycobacterium bovis*.** In *Mycobacterium bovis infection in animals and humans*. 2 edition. Edited by Thoen CO, Steele JH, Gilsdorf MJ. Ames, Iowa 50014, USA: Blackwell Publishing; 2006:18-33.

159. Hope JC, Thom ML, McCormick PA, Howard CJ: **Interaction of antigen presenting cells with mycobacteria.** *Vet Immunol Immunopathol* 2004, **100:** 187-195.

160. Stewart GR, Robertson BD, Young DB: **Tuberculosis: a problem with persistence.** *Nat Rev Microbiol* 2003, **1:** 97-105.

161. Cassidy JP, Bryson DG, Pollock JM, Evans RT, Forster F, Neill SD: **Early lesion formation in cattle experimentally infected with Mycobacterium bovis.** *Journal of Comparative Pathology* 1998, **119:** 27-44.

162. de la Rua-Domenech R, Goodchild AT, Vordermeier HM, Hewinson RG, Christiansen KH, Clifton-Hadley RS: **Ante mortem diagnosis of tuberculosis in cattle: a review of the tuberculin tests, gamma-interferon assay and other ancillary diagnostic techniques.** *Res Vet Sci* 2006, **81:** 190-210.

163. International Office of Epizootics (OIE): *Manual of Diagnostic Tests and Vaccines for Terrestrial Animals 2004*. Paris: 2006.

164. Gormley E, Doyle MB, Fitzsimons T, McGill K, Collins JD: **Diagnosis of Mycobacterium bovis infection in cattle by use of the gamma-interferon (Bovigam) assay.** *Vet Microbiol* 2006, **112:** 171-179.

165. Pollock JM, Welsh MD, McNair J: **Immune responses in bovine tuberculosis: towards new strategies for the diagnosis and control of disease.** *Vet Immunol Immunopathol* 2005, **108:** 37-43.

166. Welsh MD, Cunningham RT, Corbett DM, Girvin RM, McNair J, Skuce RA *et al.*: **Influence of pathological progression on the balance between cellular and humoral immune responses in bovine tuberculosis.** *Immunology* 2005, **114:** 101-111.

167. Plackett P, Ripper J, Corner LA, Small K, de WK, Melville L *et al.*: **An ELISA for the detection of anergic tuberculous cattle.** *Aust Vet J* 1989, **66:** 15-19.

168. Palmer MV, Waters WR: **Advances in bovine tuberculosis diagnosis and pathogenesis: what policy makers need to know.** *Vet Microbiol* 2006, **112:** 181-190.

169. Amadori M, Lyashchenko KP, Gennaro ML, Pollock JM, Zerbini I: **Use of recombinant proteins in antibody tests for bovine tuberculosis.** *Vet Microbiol* 2002, **85:** 379-389.

170. Jolley ME, Nasir MS, Surujballi OP, Romanowska A, Renteria TB, De la Mora A *et al.*: **Fluorescence polarization assay for the detection

of antibodies to Mycobacterium bovis in bovine sera. *Vet Microbiol* 2007, **120:** 113-121.

171. Krebs JR, Anderson RM, Clutton-Brock T, Donnelly CA, Frost S, Morrison WI *et al.*: **Badgers and bovine TB: conflicts between conservation and health.** *Science* 1998, **279:** 817-818.

172. Behr MA, Wilson MA, Gill WP, Salamon H, Schoolnik GK, Rane S *et al.*: **Comparative genomics of BCG vaccines by whole-genome DNA microarray.** *Science* 1999, **284:** 1520-1523.

173. Abebe F, Bjune G: **The emergence of Beijing family genotypes of Mycobacterium tuberculosis and low-level protection by bacille Calmette-Guerin (BCG) vaccines: is there a link?** *Clin Exp Immunol* 2006, **145:** 389-397.

174. Brandt L, Feino CJ, Weinreich OA, Chilima B, Hirsch P, Appelberg R *et al.*: **Failure of the Mycobacterium bovis BCG vaccine: some species of environmental mycobacteria block multiplication of BCG and induction of protective immunity to tuberculosis.** *Infect Immun* 2002, **70:** 672-678.

175. Skinner MA, Buddle BM, Wedlock DN, Keen D, de Lisle GW, Tascon RE *et al.*: **A DNA prime-Mycobacterium bovis BCG boost vaccination strategy for cattle induces protection against bovine tuberculosis.** *Infect Immun* 2003, **71:** 4901-4907.

176. McShane H, Pathan AA, Sander CR, Keating SM, Gilbert SC, Huygen K *et al.*: **Recombinant modified vaccinia virus Ankara expressing antigen 85A boosts BCG-primed and naturally acquired antimycobacterial immunity in humans.** *Nat Med* 2004, **10:** 1240-1244.

177. Vordermeier HM, Rhodes SG, Dean G, Goonetilleke N, Huygen K, Hill AV *et al.*: **Cellular immune responses induced in cattle by heterologous prime-boost vaccination using recombinant viruses and bacille Calmette-Guerin.** *Immunology* 2004, **112:** 461-470.

178. Hope JC, Thom ML, Villarreal-Ramos B, Vordermeier HM, Hewinson RG, Howard CJ: **Vaccination of neonatal calves with Mycobacterium bovis BCG induces protection against intranasal challenge with virulent M. bovis.** *Clin Exp Immunol* 2005, **139:** 48-56.

179. Haddad N, Ostyn A, Karoui C, Masselot M, Thorel MF, Hughes SL *et al.*: **Spoligotype diversity of Mycobacterium bovis strains isolated in France from 1979 to 2000.** *Journal of Clinical Microbiology* 2001, **39:** 3623-3632.

180. Dicko MS, Djitèye MA, Sangaré M: **Les systèmes de production animale au Sahel.** *Secheresse* 2006, **17:** 83-97.

181. Wilson RT. Livestock production in central Mali: Long-term studies on cattle and small ruminants in the agropastoral system. 14. 1986. Addis

Ababa, Ethiopia, ILCA.
Ref Type: Report

182. van Embden JDA, van Gorkom T, Kremer K, Jansen R, van der Zeijst BAM, Schouls LM: **Genetic variation and evolutionary origin of the direct repeat locus of Mycobacterium tuberculosis complex bacteria.** *Journal of Bacteriology* 2000, **182:** 2393-2401.

183. Streicher EM, Victor TC, van der SG, Sola C, Rastogi N, van Helden PD *et al.*: **Spoligotype signatures in the Mycobacterium tuberculosis complex.** *J Clin Microbiol* 2007, **45:** 237-240.

184. Rahim Z, Zaman K, van der Zanden AG, Mollers MJ, van SD, Raqib R *et al.*: **Assessment of population structure and major circulating phylogeographical clades of Mycobacterium tuberculosis complex in Bangladesh suggests a high prevalence of a specific subclade of ancient M. tuberculosis genotypes.** *J Clin Microbiol* 2007, **45:** 3791-3794.

185. Wiese M: *Health-vulnerability in a complex crisis situation.* Saarbrücken: Verlag für Entwicklungspolitik; 2004.

186. Aranaz A, Liebana E, Mateos A, Dominguez L, Vidal D, Domingo M *et al.*: **Spacer oligonucleotide typing of Mycobacterium bovis strains from cattle and other animals: a tool for studying epidemiology of tuberculosis.** *J Clin Microbiol* 1996, **34:** 2734-2740.

187. Zanella G, Durand B, Hars J, Moutou F, Garin-Bastuji B, Duvauchelle A *et al.*: **Mycobacterium bovis in wildlife in France.** *J Wildl Dis* 2008, **44:** 99-108.

188. Zucol F, Ammann RA, Berger C, Aebi C, Altwegg M, Niggli FK *et al.*: **Real-time quantitative broad-range PCR assay for detection of the 16S rRNA gene followed by sequencing for species identification.** *J Clin Microbiol* 2006, **44:** 2750-2759.

189. Ministère de l'Agriculture et du développement Rural de l'Algérie. Bulletin Sanitaire Vétérinaire. 2007.
Ref Type: Report

190. Bosshard PP, Abels S, Zbinden R, Bottger EC, Altwegg M: **Ribosomal DNA sequencing for identification of aerobic gram-positive rods in the clinical laboratory (an 18-month evaluation).** *Journal of Clinical Microbiology* 2003, **41:** 4134-4140.

191. Brudey K, Driscoll JR, Rigouts L, Prodinger WM, Gori A, Al-Hajoj SA *et al.*: **Mycobacterium tuberculosis complex genetic diversity: mining the fourth international spoligotyping database (SpolDB4) for classification, population genetics and epidemiology.** *BMC Microbiol* 2006, **6:** 23.

192. Prodinger WM, Brandstatter A, Naumann L, Pacciarini M, Kubica T, Boschiroli ML *et al.*: **Characterization of Mycobacterium caprae**

isolates from Europe by mycobacterial interspersed repetitive unit genotyping. *J Clin Microbiol* 2005, **43:** 4984-4992.

193. Pavlik I, Dvorska L, Bartos M, Parmova I, Melicharek I, Jesenska A *et al.*: **Molecular epidemiology of bovine tuberculosis in the Czech Republic and Slovakia in the period 1965-2001 studied by spoligotyping.** *Veterinarni Medicina* 2002, **47:** 181-194.

194. Caffrey JP: **Status of bovine tuberculosis eradication programmes in Europe.** *Vet Microbiol* 1994, **40:** 1-4.

195. Cousins D, Williams S, Liebana E, Aranaz A, Bunschoten A, Van Embden J *et al.*: **Evaluation of four DNA typing techniques in epidemiological investigations of bovine tuberculosis.** *Journal of Clinical Microbiology* 1998, **36:** 168-178.

196. Zumarraga MJ, Martin C, Samper S, Alito A, Latini O, Bigi F *et al.*: **Usefulness of spoligotyping in molecular epidemiology of Mycobacterium bovis-related infections in South America.** *Journal of Clinical Microbiology* 1999, **37:** 296-303.

197. Selander RK, Caugant DA, Ochman H, Musser JM, Gilmour MN, Whittam TS: **Methods of multilocus enzyme electrophoresis for bacterial population genetics and systematics.** *Appl Environ Microbiol* 1986, **51:** 873-884.

198. Zucol F, Ammann RA, Berger C, Aebi C, Altwegg M, Niggli FK *et al.*: **Real-time quantitative broad-range PCR assay for detection of the 16S rRNA gene followed by sequencing for species identification.** *J Clin Microbiol* 2006, **44:** 2750-2759.

199. Thoen CO: *Mycobacterium bovis infection in animals and humans*, 2nd ed edn. Ames (Iowa): Blackwell; 2006.

200. Amanfu W: **The situation of tuberculosis and tuberculosis control in animals of economic interest.** *Tuberculosis (Edinb)* 2006, **86:** 330-335.

201. Cosivi O, Meslin FX, Daborn CJ, Grange JM: **Epidemiology of Mycobacterium bovis infection in animals and humans, with particular reference to Africa.** *Rev Sci Tech* 1995, **14:** 733-746.

202. van der Zanden AG, Hoentjen AH, Heilmann FG, Weltevreden EF, Schouls LM, van Embden JD: **Simultaneous detection and strain differentiation of Mycobacterium tuberculosis complex in paraffin wax embedded tissues and in stained microscopic preparations.** *Mol Pathol* 1998, **51:** 209-214.

203. Groenen PM, Bunschoten AE, van Soolingen D, van Embden JD: **Nature of DNA polymorphism in the direct repeat cluster of Mycobacterium tuberculosis; application for strain differentiation by a novel typing method.** *Mol Microbiol* 1993, **10:** 1057-1065.

204. Fang Z, Morrison N, Watt B, Doig C, Forbes KJ: **IS6110 transposition and evolutionary scenario of the direct repeat locus in a group of closely related Mycobacterium tuberculosis strains.** *J Bacteriol* 1998, **180:** 2102-2109.

205. Warren RM, Streicher EM, Sampson SL, van der Spuy GD, Richardson M, Nguyen D *et al.*: **Microevolution of the direct repeat region of Mycobacterium tuberculosis: implications for interpretation of spoligotyping data.** *J Clin Microbiol* 2002, **40:** 4457-4465.

206. Cole ST, Brosch R, Parkhill J, Garnier T, Churcher C, Harris D *et al.*: **Deciphering the biology of Mycobacterium tuberculosis from the complete genome sequence.** *Nature* 1998, **393:** 537-544.

207. Gutacker MM, Smoot JC, Migliaccio CA, Ricklefs SM, Hua S, Cousins DV *et al.*: **Genome-wide analysis of synonymous single nucleotide polymorphisms in Mycobacterium tuberculosis complex organisms: resolution of genetic relationships among closely related microbial strains.** *Genetics* 2002, **162:** 1533-1543.

208. Tsolaki AG, Gagneux S, Pym AS, Goguet de la Salmoniere YO, Kreiswirth BN, van SD *et al.*: **Genomic deletions classify the Beijing/W strains as a distinct genetic lineage of Mycobacterium tuberculosis.** *J Clin Microbiol* 2005, **43:** 3185-3191.

209. Narayanan S, Gagneux S, Hari L, Tsolaki AG, Rajasekhar S, Narayanan PR *et al.*: **Genomic interrogation of ancestral Mycobacterium tuberculosis from south India.** *Infect Genet Evol* 2008, **8:** 474-483.

210. Brosch R, Gordon SV, Garnier T, Eiglmeier K, Frigui W, Valenti P *et al.*: **Genome plasticity of BCG and impact on vaccine efficacy.** *Proc Natl Acad Sci U S A* 2007, **104:** 5596-5601.

211. Garnier T, Eiglmeier K, Camus JC, Medina N, Mansoor H, Pryor M *et al.*: **The complete genome sequence of Mycobacterium bovis.** *Proc Natl Acad Sci U S A* 2003, **100:** 7877-7882.

212. Hanotte O, Bradley DG, Ochieng JW, Verjee Y, Hill EW, Rege JE: **African pastoralism: genetic imprints of origins and migrations.** *Science* 2002, **296:** 336-339.

213. Parra A, Larrasa J, Garcia A, Alonso JM, de Mendoza JH: **Molecular epidemiology of bovine tuberculosis in wild animals in Spain: a first approach to risk factor analysis.** *Vet Microbiol* 2005, **110:** 293-300.

214. Duarte EL, Domingos M, Amado A, Botelho A: **Spoligotype diversity of Mycobacterium bovis and Mycobacterium caprae animal isolates.** *Vet Microbiol* 2008, **130:** 415-421.

215. Dvorska L, Bartos M, Martin G, Erler W, Pavlik I: **Strategies for differentiation, identification and typing of medically important**

species of mycobacteria by molecular methods. *Veterinarni Medicina* 2001, **46:** 309-328.

216. Tadayon K, Mosavari N, Shahmoradi AH, Sadeghi F, Azarvandi A, Forbes K: **The Epidemiology of Mycobacterium bovis in Buffalo in Iran.** *J Vet Med B Infect Dis Vet Public Health* 2006, **53 Suppl 1:** 41-42.

217. Milian-Suazo F, Harris B, Diaz CA, Romero TC, Stuber T, Ojeda GA *et al.*: **Molecular epidemiology of Mycobacterium bovis: Usefulness in international trade.** *Prev Vet Med* 2008, **87:** 261-271.

218. Cobos-Marin L, Montes-Vargas J, Zumarraga M, Cataldi A, Romano MI, Estrada-Garcia I *et al.*: **Spoligotype analysis of Mycobacterium bovis isolates from Northern Mexico.** *Can J Microbiol* 2005, **51:** 996-1000.

219. Mishra GS, N'Depo AE: **[Cysticercus in animals slaughtered in the Port-Bouet abattoir (Abidjan)].** *Rev Elev Med Vet Pays Trop* 1978, **31:** 431-436.

220. Fokou G, Haller T, Zinsstag J: **[Indentification of institutional factors affecting the well-being of sedentary and nomadic populations living in the Waza-Logone flood plain along the border between Cameroon and Chad].** *Med Trop (Mars)* 2004, **64:** 464-468.

221. Kilgour V, Godfrey DG: **The influence of lorry transport on the Trypanosoma vivax infection rate in Nigerian trade cattle.** *Trop Anim Health Prod* 1978, **10:** 145-148.

222. Ben Yahmed D. Atlas du Tchad. 2006. Paris, Editions J.A.
Ref Type: Map

223. Alhaji I: **Bovine tuberculosis: a general review with special reference to Nigeria.** *The Veterinary Bulletin* 1976, 829-841.

224. Meyer CG, Scarisbrick G, Niemann S, Browne ENL, Chinbuah MA, Gyapong J *et al.*: **Pulmonary tuberculosis: Virulence of Mycobacterium africanum and relevance in HIV co-infection.** *Tuberculosis* 2008, **88:** 482-489.

225. Malaga W, Constant P, Euphrasie D, Cataldi A, Daffe M, Reyrat JM *et al.*: **Deciphering the genetic bases of the structural diversity of phenolic glycolipids in strains of the Mycobacterium tuberculosis complex.** *Journal of Biological Chemistry* 2008, **283:** 15177-15184.

226. Mostowy S, Cousins D, Behr MA: **Genomic interrogation of the dassie bacillus reveals it as a unique RD1 mutant within the Mycobacterium tuberculosis complex.** *Journal of Bacteriology* 2004, **186:** 104-109.

227. Mahairas GG, Sabo PJ, Hickey MJ, Singh DC, Stover CK: **Molecular analysis of genetic differences between Mycobacterium bovis**

BCG and virulent M-bovis. *Journal of Bacteriology* 1996, **178:** 1274-1282.

228. Brodin P, Eiglmeier K, Marmiesse M, Billault A, Garnier T, Niemann S *et al.*: **Bacterial artificial chromosome-based comparative genomic analysis identifies Mycobacterium microti as a natural ESAT-6 deletion mutant.** *Infect Immun* 2002, **70:** 5568-5578.

229. Newton SM, Smith RJ, Wilkinson KA, Nicol MP, Garton NJ, Staples KJ *et al.*: **A deletion defining a common Asian lineage of Mycobacterium tuberculosis associates with immune subversion.** *Proc Natl Acad Sci U S A* 2006, **103:** 15594-15598.

230. Candela T, Fouet A: **Poly-gamma-glutamate in bacteria.** *Mol Microbiol* 2006, **60:** 1091-1098.

231. Evans JT, Smith EG, Banerjee A, Smith RM, Dale J, Innes JA *et al.*: **Cluster of human tuberculosis caused by Mycobacterium bovis: evidence for person-to-person transmission in the UK.** *Lancet* 2007, **369:** 1270-1276.

232. van SD, Hermans PW, de Haas PE, Soll DR, van Embden JD: **Occurrence and stability of insertion sequences in Mycobacterium tuberculosis complex strains: evaluation of an insertion sequence-dependent DNA polymorphism as a tool in the epidemiology of tuberculosis.** *J Clin Microbiol* 1991, **29:** 2578-2586.

233. Garcia-Pelayo MC, Caimi KC, Inwald JK, Hinds J, Bigi F, Romano MI *et al.*: **Microarray analysis of Mycobacterium microti reveals deletion of genes encoding PE-PPE proteins and ESAT-6 family antigens.** *Tuberculosis (Edinb)* 2004, **84:** 159-166.

234. Corner LA: **The role of wild animal populations in the epidemiology of tuberculosis in domestic animals: how to assess the risk.** *Vet Microbiol* 2006, **112:** 303-312.

235. Welsh MD, Cunningham RT, Corbett DM, Girvin RM, McNair J, Skuce RA *et al.*: **Influence of pathological progression on the balance between cellular and humoral immune responses in bovine tuberculosis.** *Immunology* 2005, **114:** 101-111.

236. Jolley ME, Nasir MS: **The use of fluorescence polarization assays for the detection of infectious diseases.** *Comb Chem High Throughput Screen* 2003, **6:** 235-244.

237. Lin M, Sugden EA, Jolley ME, Stilwell K: **Modification of the Mycobacterium bovis extracellular protein MPB70 with fluorescein for rapid detection of specific serum antibodies by fluorescence polarization.** *Clin Diagn Lab Immunol* 1996, **3:** 438-443.

238. Surujballi OP, Romanowska A, Sugden EA, Turcotte C, Jolley ME: **A fluorescence polarization assay for the detection of antibodies to Mycobacterium bovis in cattle sera.** *Vet Microbiol* 2002, **87:** 149-157.

239. Waters WR, Palmer MV, Thacker TC, Bannantine JP, Vordermeier HM, Hewinson RG *et al.*: **Early antibody responses to experimental Mycobacterium bovis infection of cattle.** *Clin Vaccine Immunol* 2006, **13:** 648-654.

240. Herenda D, Chambers PG, Ettriqui A, Seneviratna P, da Silva TJP. Manual on meat inspection for developing countries. 1994. Rome, 1994, Food and Agriculture Organization (FAO) of the United Nations Rome, 1994.
Ref Type: Generic

241. Lachnik J, Ackermann B, Bohrssen A, Maass S, Diephaus C, Puncken A *et al.*: **Rapid-cycle PCR and fluorimetry for detection of mycobacteria.** *J Clin Microbiol* 2002, **40:** 3364-3373.

242. Gardner IA, Greiner M: **Receiver-operating characteristic curves and likelihood ratios: improvements over traditional methods for the evaluation and application of veterinary clinical pathology tests.** *Veterinary Clinical Pathology* 2006, **35:** 8-17.

243. Greiner M, Pfeiffer D, Smith RD: **Principles and practical application of the receiver-operating characteristic analysis for diagnostic tests.** *Prev Vet Med* 2000, **45:** 23-41.

244. Enoe C, Georgiadis MP, Johnson WO: **Estimation of sensitivity and specificity of diagnostic tests and disease prevalence when the true disease state is unknown.** *Preventive Veterinary Medicine* 2000, **45:** 61-81.

245. Corner L, Melville L, McCubbin K, Small KJ, McCormick BS, Wood PR *et al.*: **Efficiency of inspection procedures for the detection of tuberculous lesions in cattle.** *Aust Vet J* 1990, **67:** 389-392.

246. Hadorn DC, Stark KD: **Evaluation and optimization of surveillance systems for rare and emerging infectious diseases.** *Vet Res* 2008, **39:** 57.

247. Welsh MD, Cunningham RT, Corbett DM, Girvin RM, McNair J, Skuce RA *et al.*: **Influence of pathological progression on the balance between cellular and humoral immune responses in bovine tuberculosis.** *Immunology* 2005, **114:** 101-111.

248. Choi YK, Johnson WO, Collins MT, Gardner IA: **Bayesian inferences for receiver operating characteristic curves in the absence of a gold standard.** *Journal of Agricultural Biological and Environmental Statistics* 2006, **11:** 210-229.

249. Greiner M, Gardner IA: **Epidemiologic issues in the validation of veterinary diagnostic tests.** *Prev Vet Med* 2000, **45:** 3-22.

250. Varello K, Pezzolato M, Mascarino D, Ingravalle F, Caramelli M, Bozzetta E: **Comparison of histologic techniques for the diagnosis**

of bovine tuberculosis in the framework of eradication programs. *J Vet Diagn Invest* 2008, **20:** 164-169.

251. Watrelot-Virieux D, Drevon-Gaillot E, Toussaint Y, Belli P: **Comparison of three diagnostic detection methods for tuberculosis in French cattle.** *J Vet Med B Infect Dis Vet Public Health* 2006, **53:** 321-325.

252. Estrada-Chavez C, Diaz Otero F, Arriaga-Diaz C, Villegas-Sepulveda N, Perez Gonzalez R, Gonzalez Salazar D: **Agreement between PCR and conventional methods for diagnosis of bovine tuberculosis.** *Veterinaria Mexico* 2004, **35:** 225-236.

253. Moore DF, Curry JI: **Detection and Identification of Mycobacterium-Tuberculosis Directly from Sputum Sediments by Amplicor Pcr.** *Journal of Clinical Microbiology* 1995, **33:** 2686-2691.

254. *Nutrition and Health in Developing Countries*, 1st edn. Totowa, NJ, USA: Humana Press Inc.; 2001.

255. Whipple DL, Bolin CA, Miller JM: **Distribution of lesions in cattle infected with Mycobacterium bovis.** *Journal of Veterinary Diagnostic Investigation* 1996, **8:** 351-354.

256. Liebana E, Johnson L, Gough J, Durr P, Jahans K, Clifton-Hadley R *et al.*: **Pathology of naturally occurring bovine tuberculosis in England and Wales.** *Veterinary Journal* 2008, **176:** 354-360.

257. Taylor GM, Worth DR, Palmer S, Jahans K, Hewinson RG: **Rapid detection of Mycobacterium bovis DNA in cattle lymph nodes with visible lesions using PCR.** *BMC Vet Res* 2007, **3:** 12.

258. Gutierrez Cancela MM, Garcia Marin JF: **Comparison of Ziehl-Neelsen staining and immunohistochemistry for the detection of Mycobacterium bovis in bovine and caprine tuberculous lesions.** *J Comp Pathol* 1993, **109:** 361-370.

259. Reynolds D: **A review of tuberculosis science and policy in Great Britain.** *Vet Microbiol* 2006, **112:** 119-126.

260. More SJ, Good M: **The tuberculosis eradication programme in Ireland: a review of scientific and policy advances since 1988.** *Vet Microbiol* 2006, **112:** 239-251.

261. Sahraoui N, Müller B, Guetarni D, Boulahbal F, Yala D, Ouzrout R *et al.*: **Molecular characterization of Mycobacterium bovis strains isolated from cattle slaughtered at two abattoirs in Algeria.** *BMC Vet Res* 2009, **5:** 4.

262. Müller B, Hilty M, Berg S, Garcia-Pelayo MC, Dale J, Boschiroli ML *et al.*: **African 1, an epidemiologically important clonal complex of Mycobacterium bovis dominant in Mali, Nigeria, Cameroon, and Chad.** *J Bacteriol* 2009, **191:** 1951-1960.

263. Kent PT, Kubica GP. Public Health Mycobacteriology - A Guide for the Level III Laboratory. 1985. U.S.Department of Public Health and Human Services Publication, Atlanta, Georgia, 1985.
Ref Type: Report

264. Springer B, Stockman L, Teschner K, Roberts GD, Bottger EC: **Two-laboratory collaborative study on identification of mycobacteria: molecular versus phenotypic methods.** *J Clin Microbiol* 1996, **34:** 296-303.

265. Parsons LM, Brosch R, Cole ST, Somoskovi A, Loder A, Bretzel G *et al.*: **Rapid and simple approach for identification of Mycobacterium tuberculosis complex isolates by PCR-based genomic deletion analysis.** *J Clin Microbiol* 2002, **40:** 2339-2345.

266. Roring S, Scott AN, Glyn HR, Neill SD, Skuce RA: **Evaluation of variable number tandem repeat (VNTR) loci in molecular typing of Mycobacterium bovis isolates from Ireland.** *Vet Microbiol* 2004, **101:** 65-73.

267. Deshler W: **Cattle in Africa - Distribution, Types, and Problems.** *Geographical Review* 1963, **53:** 52-58.

268. Van Ert MN, Easterday WR, Huynh LY, Okinaka RT, Hugh-Jones ME, Ravel J *et al.*: **Global genetic population structure of Bacillus anthracis.** *PLoS ONE* 2007, **2:** e461.

269. Maho A, Rossano A, Hachler H, Holzer A, Schelling E, Zinsstag J *et al.*: **Antibiotic susceptibility and molecular diversity of Bacillus anthracis strains in Chad: detection of a new phylogenetic subgroup.** *J Clin Microbiol* 2006, **44:** 3422-3425.

270. Zanini MS, Moreira EC, Salas CE, Lopes MT, Barouni AS, Roxo E *et al.*: **Molecular typing of Mycobacterium bovis isolates from south-east Brazil by spoligotyping and RFLP.** *J Vet Med B Infect Dis Vet Public Health* 2005, **52:** 129-133.

271. Ostertag Rv, Kulenkampff G: *Tierseuchen und Herdenkrankheiten in Afrika.* Berlin: W. de Gruyter & co; 1941.

272. Hope JC, Thom ML, Villarreal-Ramos B, Vordermeier HM, Hewinson RG, Howard CJ: **Exposure to Mycobacterium avium induces low-level protection from Mycobacterium bovis infection but compromises diagnosis of disease in cattle.** *Clin Exp Immunol* 2005, **141:** 432-439.

273. Waters WR, Palmer MV, Thacker TC, Payeur JB, Harris NB, Minion FC *et al.*: **Immune responses to defined antigens of Mycobacterium bovis in cattle experimentally infected with Mycobacterium kansasii.** *Clin Vaccine Immunol* 2006, **13:** 611-619.

274. Michel AL: **Mycobacterium fortuitum infection interference with Mycobacterium bovis diagnostics: natural infection cases and a pilot experimental infection.** *J Vet Diagn Invest* 2008, **20:** 501-503.

275. Buddle BM, Wards BJ, Aldwell FE, Collins DM, de Lisle GW: **Influence of sensitisation to environmental mycobacteria on subsequent vaccination against bovine tuberculosis.** *Vaccine* 2002, **20:** 1126-1133.

276. Zinsstag J, Ankers P, Dempfle L, Njie M, Kaufmann J, Itty P *et al.*: **Effect of strategic gastrointestinal nematode control on growth of N'Dama cattle in Gambia.** *Veterinary Parasitology* 1997, **68:** 143-153.

277. Tanner M, Michel AL: **Investigation of the viability of M. bovis under different environmental conditions in the Kruger National Park.** *Onderstepoort J Vet Res* 1999, **66:** 185-190.

278. Bellamy R, Ruwende C, Corrah T, McAdam KP, Whittle HC, Hill AV: **Variations in the NRAMP1 gene and susceptibility to tuberculosis in West Africans.** *N Engl J Med* 1998, **338:** 640-644.

279. Coussens PM, Coussens MJ, Tooker BC, Nobis W: **Structure of the bovine natural resistance associated macrophage protein (NRAMP 1) gene and identification of a novel polymorphism.** *DNA Seq* 2004, **15:** 15-25.

280. Bonfoh B, Ankers P, Sall A, Diabaté M, Tembely S, Farah Z *et al.*: **Schéma fonctionnel de services aux petits producteurs laitiers périurbains de Bamako (Mali).** *Etudes et recherches sahéliennes* 2005, 7-25.

281. Schelling E, Bechir M, Ahmed MA, Wyss K, Randolph TF, Zinsstag J: **Human and animal vaccination delivery to remote nomadic families, Chad.** *Emerg Infect Dis* 2007, **13:** 373-379.

282. Dietrich J, Weldingh K, Andersen P: **Prospects for a novel vaccine against tuberculosis.** *Vet Microbiol* 2006, **112:** 163-169.

283. Buddle BM, Wedlock DN, Denis M: **Progress in the development of tuberculosis vaccines for cattle and wildlife.** *Vet Microbiol* 2006, **112:** 191-200.

Appendix 1:

Molecular characterization of two common Chadian cattle breeds

C. Flury[1], B. N. R. Ngandolo[2], B. Müller[3], J. Zinsstag[3] and H. N. Kadarmideen[4]

[1]Swiss Federal Institute of Technology (ETH), Zurich, Switzerland

[2]Laboratoire de Recherches Vétérinaires et Zootechniques, N'Djaména, Chad

[3]Swiss Tropical Institute, Basel, Switzerland

[4]CSIRO Livestock Industries, Rockhampton, Australia

This article has been published in:

Animal Genetic Resources Information (AGRI), Food and Agriculture Organization of the United Nations

Summary

In previous studies significant differences in *Mycobacterium bovis* infection prevalence was reported between two Chadian cattle breeds. Those findings and the established differentiation due to phenotypic characteristics suggest that the two breeds (Arab and Mbororo) are genetically different. To evaluate the genetic structure and the differences between these breeds, the genetic diversity within and between breeds was evaluated based on a total of 205 multilocus genotypes (21 microsatellite loci). All of the loci under investigation were polymorphic and the number of alleles ranged from 4-14 within the two populations. The analysis of population fixation resulted in a FST value of 0.006, further the population assignment of the individual genotypes and the exact test of population differentiation did not support the hypothesis, that the samples drawn out of the two populations are genetically different. Population admixture and sample collection are discussed as possible reasons for the rejection of the hypothesis. Finally, recommendations for sample collection in extensive systems are given.

Introduction

Mycobacterium bovis (*M. bovis*) is the causative agent of bovine tuberculosis (BTB). Bovine tuberculosis is a zoonotic disease and one question of interest is its importance in the human tuberculosis epidemic, fostered by HIV/AIDS in different parts of Africa (Ayele et al., 2004; Cosivi et al., 1998). Such investigations are extensive, as the tuberculosis epidemic and spread depend on a variety of factors such as complex interactions between different *Mycobacterium tuberculosis* complex strains, non-tuberculous mycobacteria, susceptibility of host cattle breeds, the public health status and other environmental factors. To further investigate those complexities a large project is currently running in cooperation with Laboratoire de Recherches Vétérinaires et Zootechniques de Farcha, N'Djaména, Chad; Sokoine University of Agriculture, Morogoro, Tanzania; Laboratoire Central Vétérinaire, Bamako, Mali; Ecole Inter-Etats des Sciences et de Médecine Vétérinaires, Dakar, Senegal; the Swiss Tropical Institute (STI), Basel, Switzerland and the Swiss Federal Institute of Technology (ETH), Zurich, Switzerland.

In a previous study differences between host cattle breeds regarding the prevalence of infections with *M. bovis* were reported (Hilty, 2006). In Chad as well as in Cameroon (Hilty, 2006) a higher prevalence in the Mbororo breed was found in comparison with the Arab breed and the hypothesis was, that the distinct prevalence might be due to a differential susceptibility of the two breeds. Further research on the susceptibility of different host breeds and the genetic diversity between these breeds are goals of the overall project. So far, the genetic characterizations of the samples collected at the slaughterhouses in Chad are completed and subject of the presented study.

As compared to Europe, characterization of animal genetic resources (AnGR) in Africa receives less attention. In a study on the Kuri cattle breed mentioned in the country report of Chad (FAO, 2007b), no molecular characterization of Chadian cattle breeds was reported. However, adequate characterization of AnGR is a prerequisite for successful management programs and for informed decision making in national livestock development (FAO, 2007a). Even if the two breeds Mbororo and Arab are not under risk of extinction (derived from FAO, 2007c) the data collected at the slaughterhouses in Chad is expected to be worth for a detailed analysis of molecular aspects of them.

The aim of this study was the molecular characterization of the two breeds including the assessment of genetic diversity within and between populations. Such a characterization is of primary interest regarding the differences in BTB prevalence of the two breeds but also in respect to the description of indigenous African cattle breeds and African cattle husbandry systems.

Material and Methods

Breeds

The genotyped animals belong to the two breeds Mbororo and Arab. All of them were kept in a long distance transhumant system by pastoralists, thereby passed the border between Chad and Central African Republic and spent the dry season in the Central African Republic. The transhumant system is the main cattle production system in Chad. Seventy five percent of the national herds are kept by pastoralists and almost 50% of Chadian export revenues are generated within this system (FAO, 2007b).

The Mbororo cattle, also known as Red Fulani, belong to the subgroup Fulani of the West African Zebu cattle. In Chad a population size of 300'000 heads was reported in year 1992 (derived from FAO, 2007c). This breed has long, lyre-shaped horns and a thoracic, sometimes intermediate hump (derived from FAO, 2007c) (figure 12). The lactation yield is poor with 2 kg of milk per day at the peak of lactation (FAO, 2007b). The carcass dressing out is low (40% - 42%), but FAO (2007b) reported the good quality of the breed's hides for leather production. The breed is robust and adapted to different climates, i.e. the breed is kept in dry as well as humid regions of Chad (FAO, 2007b).

The Arab zebu (or Shewa) has a well developed dewlap and short horns (Zibrowski, 1997). Coat color is red – maroon in the Sahel-zone and predominantly white in the south-east and west (FAO, 2007b). Figure 13 shows some Arab animals from Chad before slaughter. Milk yield per lactation varies from a minimum of 454 kg to a maximum of 1814 kg and lactation length varies from 240 to 396 days (derived from DAGRIS, 2007). Beside the totally deserted regions, the breed is kept in all other regions of Chad. It is estimated, that 75% to 90% of Chadian cattle belong to this breed (FAO, 2007b). A population size of 4'902'000 heads was reported in year 1992 (derived from FAO, 2007c).

Figure 12. Mbororo cattle at the slaughterhouse in Chad (photo Ngandolo B.N.R.).

Figure 13. Arab cattle at the slaughterhouse in Chad (photo Ngandolo B.N.R.).

Genotyping

Blood samples were from animals before slaughter at three different abattoirs in southern Chad. Additionally, information about the breed, the age, sex, the transhumance system, borders passed, the residence during dry season and the location of the slaughter house of each individual was recorded. The age structure and gender of the sampled individuals is subject of table 19.

Blood was allowed to clot, transported on ice to the Laboratoire de Recherches Vétérinaires et Zootechniques in Farcha and stored at -80°C until further processing. DNA was extracted using the QIAamp® DNA Blood Mini Kit (QIAGEN, Cat. No. 51106) from clotted blood corresponding to 238 individual animals. Handling was carried out according to the Blood and Body Fluid Spin protocol (derived from Qiagen, 2007). DNA was transported to Europe where genotyping was conducted by Van Haeringen Laboratories, Wageningen, Netherlands. All microsatellites were chosen from the FAO-list (FAO, 2004).

A remarkable degradation of the DNA was observed over time. This problem caused a high fraction of missing genotypes, especially for the most recent genotyped multiplexes. Markers with individuals with missing information for seven and more markers were omitted from further analysis. Finally, 205 genotypes (131 Arab and 74 Mbororo) for 21 microsatellites were included for statistical analysis.

Breed	sex	N	age (mean)	Number of animals per age class (years)										
				1	2	3	4	5	6	7	8	9	10	11
Arab	male	34	4.206	6	4	6	2	6	3	3	2	2	0	0
	female	97	6.701	2	1	3	6	6	20	26	21	9	1	2
Mbororo	male	38	3.079	2	18	7	4	4	0	2	1	0	0	0
	female	36	5.611	1	5	3	2	2	6	9	7	1	0	0

Table 19. Age structure and average age of the sampled individuals (grouped by sex and breed).

Statistical analysis

For the statistical investigations the packages ARLEQUIN 3.01 (Excoffier et al., 2005), STRUCTURE 2.1 (Pritchard et al., 2000) and FSTAT 2.9.3.2 (Goudet,

1995) were applied. Deviation from Hardy-Weinberg- Equilibrium (HWE) was tested for each locus in each population using ARLEQUIN (number of steps in MCMC =100´000). The significance level was set to p-value <0.001.

FAO- markers are assumed to be selectively neutral and to segregate independently from other loci (FAO, 2004). In ARLEQUIN a likelihood ratio test of linkage disequilibrium is implemented for genotypic data with unknown gametic phase. This test was conducted on the data setting the number of permutations to 10´000 and the significance level to 0.05.

The number of alleles per locus, the average number of alleles per breed, the observed and expected heterozygosity per locus and breed were calculated as indicators for the genetic variability within the two breeds. The relevant results were part of the testing on HWE with ARLEQUIN. Further breed specific alleles (i.e. private alleles) were counted.

FSTAT (Goudet, 1995) was used for the assessment of Wrights fixation indices and the respective standard errors. Further, the computations given in ARLEQUIN to conduct population comparisons and population differentiation were conducted. Additionally genotype assignment was derived with this package.

Clustering analysis was conducted with STRUCTURE 2.1 (Pritchard et al., 2000). The length of burning period for the MCMC was set to 10000 with 100000 replications after burning. The number of clusters was varied from 2 to 5.

Results

Information content of markers and genetic variability within populations

Table 20 gives an overview of the genotyped markers, the number of individuals with a genotype (N), the number of observed alleles, the fraction of animals with missing genotypes, the observed heterozygosity and the expected heterozygosity and the respective p-value for HWE-testing for the two populations Arab and Mbororo, separately.

Genetic diversity between populations and cluster analysis

The total degree of population subdivision according to Weir and Cockerham (1984) was found to be:

FIT= 0.042 (± 0.008)
FST= 0.006 (± 0.002)
FIS= 0.037 (± 0.008).

Figure 14 shows the results for the genotype assignment implemented in ARLEQUIN. The program calculates the log-likelihood of each genotype under the assumption that it belongs to the respective population.

The results of the clustering analysis assuming two clusters are given in figure 15. The number of clusters (k) investigated is user defined. The k resulting in the highest logarithmic probability is seen as the most probable number of subpopulations. For our data the highest log-likelihood was found for k=2.

	Arab (131)						Mbororo (74)					
Marker	N	alleles	missing	obs_het	exp_het	p-value	N	alleles	missing	obs_het	exp_het	p-value
BM1818	129	9	1.50%	0.814	0.813	0.894	69	9	6.80%	0.783	0.844	0.238
BM1824	131	7	0.00%	0.672	0.746	0.003	74	4	0.00%	0.703	0.743	0.91
BM2113	131	8	0.00%	0.771	0.827	0.809	74	9	0.00%	0.824	0.822	0.734
CSRM60	131	10	0.00%	0.595	0.568	0.677	74	8	0.00%	0.541	0.61	0.202
CSSM66	131	11	0.00%	0.771	0.832	0.322	74	9	0.00%	0.757	0.838	0.006
ETH10	129	8	1.50%	0.798	0.769	0.382	74	8	0.00%	0.797	0.803	0.237
ETH225	129	9	1.50%	0.605	0.635	0.725	74	8	0.00%	0.689	0.714	0.341
ETH3	131	8	0.00%	0.618	0.6	0.504	74	7	0.00%	0.514	0.553	0.842
HAUT27	113	7	13.70%	0.664	0.744	0.374	66	7	10.80%	0.727	0.746	0.797
ILSTS006	126	10	3.80%	0.651	0.687	0.631	70	10	5.40%	0.786	0.75	0.654
INRA23	130	12	0.80%	0.708	0.745	0.526	74	10	0.00%	0.595	0.665	0.281
SPS115	131	7	0.00%	0.496	0.497	0.81	74	7	0.00%	0.338	0.348	0.766
TGLA122	126	14	3.80%	0.683	0.716	0.154	72	11	2.70%	0.722	0.704	0.686
TGLA126	131	8	0.00%	0.786	0.755	0.756	74	8	0.00%	0.716	0.756	0.055
TGLA227	131	10	0.00%	0.618	0.621	0.287	74	11	0.00%	0.5	0.572	0.223
TGLA53*	93	16	29.00%	0.763	0.787	0.747	60	15	18.90%	0.567	0.768	0.002
ETH152	131	6	0.00%	0.511	0.525	0.827	73	5	1.40%	0.507	0.527	0.815
ETH185	129	14	1.50%	0.806	0.823	0.478	74	11	0.00%	0.662	0.757	0.282
HEL5*	96	7	26.70%	0.573	0.77	0	52	6	29.70%	0.5	0.751	0
ILSTS005	121	6	7.60%	0.76	0.792	0.386	71	5	4.10%	0.732	0.752	0.895
INRA32	126	10	3.80%	0.714	0.826	0.033	72	10	2.70%	0.806	0.815	0.949
INRA35	128	8	2.30%	0.5	0.534	0.091	71	7	4.10%	0.577	0.671	0.004
MM12	131	14	0.00%	0.832	0.839	0.962	74	11	0.00%	0.838	0.859	0.489
Mean**		9.3		0.684	0.709			8.3		0.672	0.707	
SD**		2.5		0.106	0.113			2.1		0.133	0.127	

* excluded
** after exlcusion of markers TGLA53 and HEI

Table 20. Overview of genotyping parameters. Number of genotypes (N), number of alleles, fraction of missing genotypes observed heterozygosity, expected heterozygosity and p-value for HWE-testing for the Arab sample and the Mbororo sample, respectively.

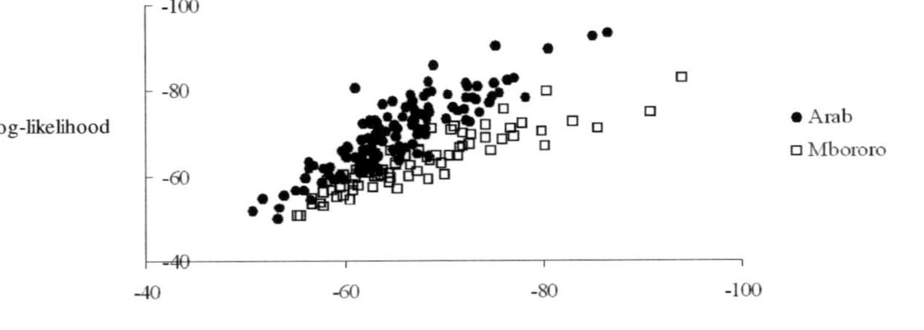

Figure 14. Log-likelihood of each individual's multilocus genotype in the population sample Arab and Mbororo, respectively, assuming that it comes from this population.

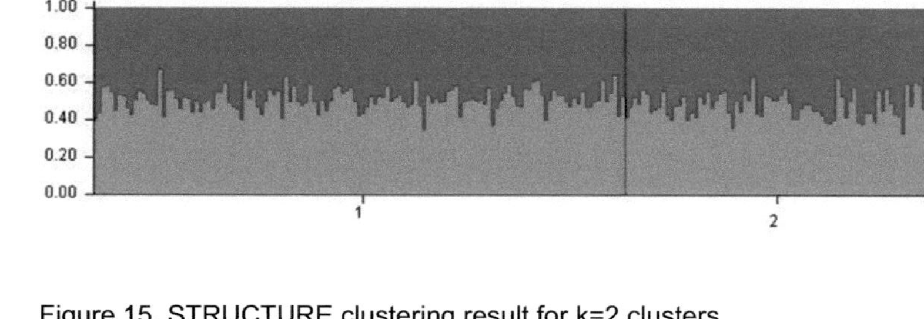

Figure 15. STRUCTURE clustering result for k=2 clusters.

Discussion

Information content of markers and genetic variability within populations

Marker HEL5 highly significant deviates from HWE and is therefore excluded from further analysis. Further TGLA53 was omitted, as its fraction of missing genotypes was above 20%. After exclusion of the above mentioned markers, 205 individual genotypes for totally 21 microsatellites remained for further analysis (table 20).

Testing on linkage disequilibrium revealed that for each population three pairs of loci do not segregate independently (p<0.001) (results not shown). However, as all of the markers in linkage disequilibrium are mapped to different chromosomes, the markers are informative regarding diversity studies and are not excluded from further analysis (Peter, 2005).

The number of alleles per locus ranged from 4 up to 14. The minimum was found in the Mbororo sample at the loci BM1824, the maximum at the three loci TGLA122, ETH185 and MM12 of the Arab sample (table 20). These findings show that the two populations are polymorphic for all of the 21 loci under investigation. The chosen loci all fulfill the rule of thumb given by FAO, that markers for diversity studies should segregate with at least 4 alleles per population (FAO, 2004). The mean number of alleles was 9.3 (± 2.5) for the genotypes belonging to the breed Arab and 8.3 (± 2.1) for the genotypes belonging to the Mbororo breed, averaging 8.8 (± 2.3) (table 20) for the total sample.

28 alleles at 13 loci out of the 203 alleles were found to be so called private alleles (results not shown). A private allele is defined as an allele found in one population but in no other (Woolliams and Toro, 2007). In our study the highest frequency of a private allele was 2.8% only. Thus, their influence on differences in the allelic frequencies between populations is expected to be low.

The average observed heterozygosity was found to be 0.684 (± 0.106) in Arab and 0.672 (± 0.133) in Mbororo, respectively. The average expected heterozygosity was 0.709 (± 0.113) for population Arab, and 0.707 (± 0.127) for the population Mbororo (table 20). The mean number of alleles per locus and the expected heterozygosity are seen as informative measures for the assessment of genetic diversity within populations (Hanotte and Jianlin, 2005; Toro and Caballero, 2004). The mean number of alleles per locus found in the present study is lower than the 11.5 alleles per microsatellite locus observed by Ibeagha-Awemu et al. (2004) in west/central African cattle breeds. The expected heterozygosity for the nine *Bos indicus* breeds investigated by Ibeagha-Awemu et al. (2004) ranged from 0.703 – 0.744. Our estimates correspond with the lower end of this range.

Generally, it has to be questioned if the samples drawn for our study reflect random samples from the Mbororo and Arab breed or not. The number of animals sampled is adequate, however, the animals were all kept in one region of southern Chad and the size of the two samples was not equal. A balanced affiliation of both sexes is not given for the Arab sample (table 19). Further, the animals from a pastoralist system arriving at abattoir do not necessarily cover all age classes of a population. This expectation is supported with table 19. For both breeds the average age of the sampled cows is about 2.5 years higher than the average age of the sampled bulls (table 19). Considering bulls, animals from the older age classes (> 6 years) are under represented in both breeds, indicating that the majority of bulls are slaughtered in younger years (table 19). Older animals might have undergone selection as they had to survive dry season, long treks, disease pressure and other forces arising within this system. Due to these various factors, the assumption of two random samples can not be warranted.

Genetic diversity between populations and cluster analysis

The FST indicates that the genetic diversity between the two samples is very low. A high proportion of the FIT is accounted for by the within-heterozygote deficiency (FIS). The low FST is seen as a first incidence, that it might be hard to elaborate genetic differences between the samples of Mbororo and Arab cattle.

The distributions of the log-likelihoods for the genotype assignment shown in figure 14 overlap to a certain amount. Again it is not possible to clearly distinguish between the two populations. This result was further confirmed with the exact test of population differentiation implemented in ARLEQUIN (results not shown). The differentiation test between all samples revealed in p-value > 0.05, i.e. based on the genotypic information, the two populations do not significantly differ.

The algorithm implemented in STRUCTURE (Pritchard et al., 2000) constructs genetic clusters from a collection of individual multilocus genotypes. Therefore the fraction of each individuals genotype that belongs to each cluster is estimated (Rosenberg et al., 2001). It identifies sub-populations which differ in their allele frequencies.

The bars in figure 15 show, that for none of the 205 individuals the genome can be clearly assigned to the Arab cluster or the Mbororo cluster. Furthermore, no relation between the participation of an individual's genome fraction and its initially assigned population (x-axis in figure 15) was found.

Rosenberg et al. (2001) showed that the power of clustering depends on the variability of markers, the number of markers and the number of individuals genotyped. For less diverged populations they propose to genotype more than 12-15 markers for 15-20 individuals of the hypothetical populations to get accurate clustering results. For our data those recommendations are fulfilled. Therefore the clustering results further support, that the samples under investigation do not reflect genetically different populations.

Before slaughtering the sampled individuals were phenotypically assigned to the two breeds Mbororo and Arab. Even if relying on different individuals sampled, the reported differences in BTB prevalence between the two breeds (Hilty, 2006) lead to the hypothesis, that genetic differences exist and might become obvious investigating the molecular diversity. However, the analysis of the samples investigated here and the chosen microsatellites do not support this hypothesis. Those findings are somewhat unexpected. They might be explained with effects regarding the sampling of animals kept in transhumance systems. Unfortunately, no data about the herd affiliation was available. As already mentioned above different age structures were observed between sexes. There is a certain chance that "old" female individuals (5 to 8 years) are the ones that survived for example BTB infection and are therefore "overrepresented" in both samples. Such sampling effects can result in diminished differences between breeds.

Mbororo and Arab animals are kept by nomadic pastoralists of two different ethnic groups, named according to their cattle breeds. This connection appears to be rather loose and both groups often keep Arab and Mbororo cattle intermixed in their herds (Dr. C. Diguimbaye-Djaïbe and B.N.R. Ngandolo, personal communications). Another possibility is that migration of animals between herds and breeds occur. These aspects support the rejection of the hypothesis due to population admixture. Admixture between populations homogenizes allele frequencies between populations. Therefore, the exploration of differences in allele frequencies between admixed populations does not lead to significant testing results. This conclusion is further supported by the Country Report of Chad (FAO, 2007b) which records, that important admixture between Arab and Mbororo exists.

Based on our study, we fully support the statement, that sample collection is the most important step in any diversity study (FAO, 2007a). In extensive production system the lack of pedigree information (Eding and Meuwissen, 2001; Ruane, 1999) may hamper the collection of representative samples. To overcome this difficulty a well planned data collection and the collection of additional information like herd affiliation, records of geographical coordinates and photo documentation of sampling sites, animals and flocks. are highly recommended (FAO, 2007a). Otherwise, the interpretation of genotyping results and statistical analysis become hard and loose the explanatory power.

Conclusions

Considering phenotypes solely, one would have presumed two different breeds. However, our study does not confirm genetic differences between the two samples. Here, the potential of genetic characterization studies in extensive systems becomes obvious. The presented results increase information about cattle breeds kept in pastoralists system and support that regular admixture between the two breeds occurs.

Collecting sample at slaughterhouses for semi-feral populations seems promising in comparison with the complex collection of field samples. Even though, careful sample collection procedure remains the most important step. In this context the need for supplementary information (description of the breeds, herd information, information about herd management etc.) is underlined. For this purpose, the pastoralists arriving at slaughterhouse might be asked to fill in a questionnaire. Future research also requires investigations on cattle husbandry and herding practices of African pastoral communities where very little information is available. No detailed information about the influence of nongenetic factors on differences in disease prevalence (i.e. BTB) between breeds is available.

Increased information about the genetic composition of breeds as well as their production system allows for better understanding of pastoralists system in general and of specific threats - such as zoonotic diseases – arising within such systems.

Acknowledgements

This study is part of an overall SNF project. SNF is acknowledged for financial support. Stefan Rieder and two independent referees are acknowledged for their comments on earlier versions of this manuscript.

References

Ayele, W. Y., Neill, S. D., Zinsstag, J., Weiss, M. G., & Pavlik, I. (2004). Bovine tuberculosis: an old disease but a new threat to Africa. Int. J. Tuberc. Lung Dis. 8, 924-937.

Cosivi, O., Grange, J. M., Daborn, C. J., Raviglione, M. C., Fujikura, T., Cousins, D., Robinson, R. A., Huchzermeyer, H. F., de, K., I, & Meslin, F. X. (1998). Zoonotic tuberculosis due to Mycobacterium bovis in developing countries. Emerg. Infect. Dis. 4, 59-70.

DAGRIS, 2007. Domestic Animal Genetic Resources Information System (DAGRIS). (eds. J.E.O. Rege, O. Hanotte, Y. Mamo, B. Asrat and T. Dessie). http://dagris.ilri.cgiar.org/, International Livestock Research Institute, Addis Ababa, Ethiopia.

Eding, H., & T. H. E. Meuwissen. 2001. Marker-based estimates of between and within population kinships for the conservation of genetic diversity. Journal of Animal Breeding and Genetics 118: 141-159.

Excoffier, L., G. Laval, & S. Schneider. 2005. Arlequin (version 3.0): An integrated software package for population genetics data analysis Evolutionary Bioinformatics Online.

FAO. 2007a. The state of the world's animal genetic resources for food and agriculture, B. Rischkowsky and D. Pilling (eds.), Rome, Italy, pp 511.

FAO. 2007b. The state of the world's animal genetic resources for food and agriculture - annex: Country report chad. B. Rischkowsky and D. Pilling (eds.), Rome, Italy, pp 511.

FAO. 2007c. Domestic animal diversity information system (DAD-IS). http://dad.fao.org/, FAO, Rome, Italy.

FAO. 2004. Secondary guidelines for development of national farm animal genetic resources management plans, Rome, Italy, pp 55.

Goudet, J. 1995. Fstat (version 1.2): A computer program to calculate f-statistics. Journal of Heredity 86: 485-486.

Hanotte, O., & H. Jianlin. 2005. Genetic characterization of livestock populations and its use in conservation decision-making. In: The Role of Biotechnology for the characterization and conservation of crop, forestry, animal and fishery genetic resources. FAO Workshop, Turin, Italy

Hilty, M. 2006. Molecular epidemiology of mycobacteria: Development and refinement of innovative molecular typing tools to study mycobacterial infections, Universität Basel, Basel, Switzerland, pp 157.

Ibeagha-Awemu, E. M., O. C. Jann, C. Weimann, & G. Erhardt. 2004. Genetic diversity, introgression and relationships among west/central african cattle breeds. Genetics Selection Evolution 36: 673-690.

Peter, C. 2005. Molekulargenetische Charakterisierung von Schafrassen Europas und des Nahen Ostens auf der Basis von Mikrosatelliten, Justus-Liebig-Universität, Giessen, Germany, pp 160.

Pritchard, J. K., M. Stephens, & P. Donnelly. 2000. Inference of population structure using multilocus genotype data. Genetics 155: 945-959.

Qiagen. 2007. Qiagen: sample and assay technologies. http://www1.qiagen.com/literature/handbooks/literature.aspx?id=1000190 .

Rosenberg, N. A., Burke, T., Elo, K., Feldman, M. W., Freidlin, P. J., Groenen, M. A. M., Hillel, J., Mäki-Tanila, A., Tixier-Boichard, M., Vignal, A., Wimmers, K. & S. Weigend. 2001. Empirical evaluation of genetic clustering methods using multilocus genotypes from 20 chicken breeds. Genetics 159: 699-713.

Ruane, J. 1999. A critical review of the value of genetic distance studies in conservation of animal genetic resources. Journal of Animal Breeding and Genetics 116: 317-323.

Toro, M., & Caballero. 2004. Characterisation and conservation of genetic diversity between breeds. In: Proceedings 55th EAAP Annual Meeting, Bled, Slovenia.

Weir, B., & C. Cockerham. 1984. Estimating f-statistics for the analysis of population structure. Evolution 38: 12.

Woolliams, J., & M. Toro. 2007. Chapter 3. What is genetic diversity? In: Utilisation and conservation of farm animal genetic resources. Wageningen Academic Publishers, Netherlands, 55-74.

Zibrowski, D. 1997. Atlas d'élevage du bassin du lac tschad / livestock atlas of the lake chad bassin. CIRAD-EMVT, Wageningen, Netherlands, 79-80.

Appendix 2:

Supplementary data for: "Molecular characterization of Mycobacterium bovis strains isolated from cattle slaughtered at two abattoirs in Algeria"

Sample name	Spoligotype	ETR-A	ETR-B	ETR-C	ETR-D	ETR-E
4 'G	SB0121	4	3	5	3*	3
6	SB0837	6	4	3	4*	3
8	SB0121	6	4	3	4*	3
9 'G	SB0860	5	3	5	4*	3
9	SB0120	5	5	5	4*	4
10	SB0120	5	5	2	4*	3
11 '	SB1448	7	4	5	4*	3
13	SB0120	5	5	2	4*	3
14 '	SB0121	6	4	3	4*	3
14	SB0120	4	5	3	4*	3
16 '	SB1450	4	5	5	4*	3
16	SB0120	4	5	5	2*	3
17 'C	SB0121	6	4	3	4*	3
17	SB0121	6	4	3	4*	3
18	SB0121	6	4	3	4*	3
20 '	SB0121	6	4	3	4*	3
21 '	SB0120	5	3	5	4*	3
21 p	SB0941	6	7	3	4*	3
22 '	SB0120	5	3	5	4*	3
22	SB0120	5	4	5	4*	3
23	SB0120	4	5	5	4*	3
26 '	SB0134	3	5	5	5*	3
26	SB0941	6	5	3	4*	3
27	SB0120	4	5	5	4*	3
28	SB0828	5	5	5	4*	4
29 'G	SB0134	7	4	5	4*	3
31 'C	SB0120	3	2	5	4*	3
32 'p	SB1200	4	6	5	4*	3
33	SB0850	5	5	5	4*	3
34 'G	SB0120	4	5	5	4*	3
36 G	SB0134	4	4	5	3*	4
38	SB0120	3	5	5	4*	4
39 '	SB0121	6	4	3	4*	3
44 '	SB0121	6	4	3	4*	3
44 p	SB1086	3	3	5	4*	3
45	SB0120	7	5	4	3*	3
46 '	SB0120	4	5	3	4*	3
47 'T	SB0331	3	3	5	4*	3
47	SB0867	4	5	5	4*	3
49 '	SB0121	6	4	3	4*	3
49	SB0120	4	5	5	2*	3
50 'T	SB0121	6	4	3	4*	3
51 '	SB0121	6	4	3	4*	3
56 F	SB0120	5	5	5	4*	3
57 '	SB0831	5	5	5	4*	3

Sample name	Spoligotype	ETR-A	ETR-B	ETR-C	ETR-D	ETR-E
57	SB0121	6	4	4	4*	3
61 '	SB0120	4	5	5	3*	3
62 p	SB0822	3	5	5	4*	3
64	SB0120	4	4	5	4*	3
65 '	SB0120	5	5	5	4*	4
65	SB0120	4	5	2	4*	3
66 '	SB0822	5	7	5	4*	3
68 '	SB0120	4	5	5	4*	3
72 '	SB1451	6	4	3	4*	5
72 p	SB0873	5	6	5	4*	3
74 '	SB0941	6	5	3	4*	3
74	SB0120	3	4	5	3*	3
75 p	SB1452	6	4	5	3*	3
77	SB1449	6	4	5	4*	2
79	SB0121	5	3	5	4*	3
80 '	SB0120	5	4	4	4*	3
81 p	SB0941	6	5	3	4*	3
84 p	SB0941	6	5	3	4*	3
85 '	SB0121	6	4	3	4*	3
85 p	SB0941	6	5	3	4*	3
86 '	SB0120	5	5	2	4*	3
87 '	SB0120	4	4	5	4*	3
88 '	SB0121	6	4	3	4*	3
89 '	SB0162	5	5	5	4*	3
89	SB0121	6	4	3	4*	3
90 '	SB0120	4	5	5	3*	3
90 p	SB0121	6	4	3	4*	3
92 p	SB0121	6	4	3	4*	3
93 '	SB0132	6	4	3	4*	3
93 G	SB0120	4	8	5	4*	3
95 '	SB1447	6	5	5	4*	3
97 p	SB0120	5	5	2	4*	3
98	SB0120	4	5	3	4*	3
102 '	SB1450	6	5	5	4*	3
105 '	SB0120	4	5	5	4*	3
106 '	SB0134	7	4	5	4*	3
107 '	SB0134	7	4	5	4*	3
108 '	SB0860	5	4	5	4*	3
110 '	SB0120	4	7	3	4*	3
111 '	SB0120	5	6	5	4*	3
114 '	SB0850	5	4	5	4*	3
117 '	SB0120	4	5	5	4*	3
118 '	SB0120	4	5	5	3*	3
141 '	SB0134	7	5	3	5*	3

Supplementary data 1. Original sample name, spoligotype number according to www.Mbovis.org and ETR A-E typing results for all 89 MTBC strains isolated.

Appendix 3:

Supplementary data for: "African 1; an epidemiologically important clonal complex of *Mycobacterium bovis* dominant in Mali, Nigeria, Cameroon and Chad"

	Lab. ID	Strain ID	Spoligotype	VNTR	RDAf1	Spacer 30	Accesion	Where isolated	When isolated (Month, Year)
Chad	1	90 TUB/G	SB0944	ND	deleted	absent		N'Djaména Abattoir, southwestern Chad	07.08.2000
n = 65	2	106PM/P	SB0944	3 6 5 4* 3 3.1	deleted	absent		N'Djaména Abattoir, southwestern Chad	22.08.2000
	3	117PM/P	SB0951	7 5 5 4* 3 3.1	deleted	absent		N'Djaména Abattoir, southwestern Chad	09.09.2000
	4	120PM/P	SB1025	4 3 4 4* 3 3.1	deleted	absent		N'Djaména Abattoir, southwestern Chad	09.09.2000
	5	215PM/P	SB0951	X X X X 3 X	deleted	absent		N'Djaména Abattoir, southwestern Chad	09.09.2000
	37	234GG/G	SB1418	3 6 5 4* 3 3.1	deleted	absent		N'Djaména Abattoir, southwestern Chad	02.08.2001
micro-array	38	86PM/P	SB1103	7 5 5 4* 3 3.1	intact	absent	seq Af1 intact	N'Djaména Abattoir, southwestern Chad	04.08.2000
	39	453GG/P	SB0952	5 5 5 4* 3 3.1	deleted	absent		N'Djaména Abattoir, southwestern Chad	27.07.2002
	40	465GG/P	SB1025	ND	deleted	absent		N'Djaména Abattoir, southwestern Chad	25.07.2002
	41	500GG/P	SB1027	X X X X 3 X	deleted	absent		N'Djaména Abattoir, southwestern Chad	05.08.2002
	42	579GG/P	SB0951	2 6 5 4* 3 3.1	deleted	absent		N'Djaména Abattoir, southwestern Chad	23.08.2002
	43	587GG/P	SB1025	4 3 4 4* 3 3.1	deleted	absent		N'Djaména Abattoir, southwestern Chad	24.08.2002
	44	94GG/P	SB1027	4 6 4 4* 3 3.1	deleted	absent		N'Djaména Abattoir, southwestern Chad	09.08.2000
	45	241GG/P	SB0944	3 5 5 4* 3 3.1	deleted	absent	EU887550	N'Djaména Abattoir, southwestern Chad	15.10.2001
	46	466*GG/P	SB0944	ND	deleted	absent		N'Djaména Abattoir, southwestern Chad	25.07.2002
	47	474GG/P	SB1027	X X X X 3 2.1	deleted	absent		N'Djaména Abattoir, southwestern Chad	29.07.2002
	48	506PM/G	SB1099	X X X 4* 3 X	deleted	absent		N'Djaména Abattoir, southwestern Chad	07.08.2002
	49	564GG/P	SB0944	5 5 3 4* 3 3.1	deleted	absent		N'Djaména Abattoir, southwestern Chad	22.08.2002
	50	572GG/P	SB1098	4 5 5 4* 3 3.1	deleted	absent	EU887551	N'Djaména Abattoir, southwestern Chad	22.08.2002
micro-array	51	111GG/P	SB0952	ND	deleted	absent		N'Djaména Abattoir, southwestern Chad	01.09.2000
	52	245GG/P	SB0944	4 7 5 4* 3 3.1	deleted	absent		N'Djaména Abattoir, southwestern Chad	05.10.2001
	53	468GG/G	SB0952	5 5 5 4* 3 2.1	deleted	absent		N'Djaména Abattoir, southwestern Chad	26.07.2002
micro-array	54	486GG/P	SB0944	4 8 5 4* 3 3.1	deleted	absent	EU887558	N'Djaména Abattoir, southwestern Chad	01.08.2002
	55	510PM/G	SB1025	X 3 4 4* 3 X	deleted	absent	EU887556	N'Djaména Abattoir, southwestern Chad	07.08.2002
	56	573GG/P	SB0944	4 7* 5 4* 3 3.1	deleted	absent		N'Djaména Abattoir, southwestern Chad	22.08.2002
	57	570GG/P	SB1027	4 4 5 4* 3 2.1	deleted	absent		N'Djaména Abattoir, southwestern Chad	22.08.2002
	58	216PM/P	SB0944	3 5 3 4* 3 3.1	deleted	absent		N'Djaména Abattoir, southwestern Chad	07.08.2001
	60	488GG/P	SB1100	4 6 5 4* 3 3.1	deleted	absent		N'Djaména Abattoir, southwestern Chad	01.08.2002
	61	600GG/P	SB1098	4 5 5 4* 3 3.1	deleted	absent		N'Djaména Abattoir, southwestern Chad	26.08.2002
	62	605GG/P	SB0944	4 3 5 4* 3 3.1	deleted	absent		N'Djaména Abattoir, southwestern Chad	26.08.2002
	63	614*GG/P	SB1025	5 5 4 4* 3 3.1	deleted	absent		N'Djaména Abattoir, southwestern Chad	26.08.2002

	64	617*GG/P	SB0951	4 X 5 4* 3 3.1	deleted	absent		N'Djaména Abattoir, southwestern Chad	26.08.2002
	65	630*PM/G	SB0951	4 6 5 4* 3 3.1	deleted	absent		N'Djaména Abattoir, southwestern Chad	28.08.2002
	66	638GG/P	SB0951	4 6 5 4* 3 3.1	deleted	absent		N'Djaména Abattoir, southwestern Chad	28.08.2002
	67	646GG/P	SB1101	ND	deleted	absent		N'Djaména Abattoir, southwestern Chad	27.08.2002
	69	653PM/G	SB1025	4 3 4 4* 3 3.1	deleted	absent		N'Djaména Abattoir, southwestern Chad	29.08.2002
	70	659*GG/G	SB1025	5 5 4 4* 3 3.1	deleted	absent		N'Djaména Abattoir, southwestern Chad	29.08.2002
	71	660GG/G	SB0951	4 6 5 4* 3 3.1	deleted	absent		N'Djaména Abattoir, southwestern Chad	29.08.2002
	72	220PM/P	SB0944	4 5 6 4* 2 3.1	deleted	absent		N'Djaména Abattoir, southwestern Chad	25.08.2001
	73	221GG/P	SB0944	4 5 5 2 3 3.1	deleted	absent		N'Djaména Abattoir, southwestern Chad	25.08.2001
	75	449GG/P	SB0944	3 6 5 4* 3 3.1	deleted	absent		N'Djaména Abattoir, southwestern Chad	22.07.2002
	76	450GG/P	SB0944	X X 5 4* 3 X	deleted	absent		N'Djaména Abattoir, southwestern Chad	22.07.2002
	77	452GG/P	SB1103	7 5 5 4* 3 3.1	intact	absent		N'Djaména Abattoir, southwestern Chad	22.07.2002
	78	471PM/6	SB0944	X X 5 X 3 X	deleted	absent		N'Djaména Abattoir, southwestern Chad	29.07.2002
	79	472PM/P	SB0328	5 5 3 4* 3 3.1	deleted	absent		N'Djaména Abattoir, southwestern Chad	29.07.2002
	80	473GG/P	SB0944	4 8 5 4* 3 3.1	deleted	absent		N'Djaména Abattoir, southwestern Chad	29.07.2002
	81	491PM/G	SB0944	4 7 6 4* 3 3.1	deleted	absent	EU887549	N'Djaména Abattoir, southwestern Chad	02.08.2002
	82	493GG/P	SB0944	4 6 5 4* 3 3.1	deleted	absent		N'Djaména Abattoir, southwestern Chad	03.08.2002
	83	496GG/P	SB1101	4 X 5 4* 3 3.1	deleted	absent		N'Djaména Abattoir, southwestern Chad	05.08.2002
	84	511GG/G	SB1102	3 X X 4* 3 3.1	intact	present		N'Djaména Abattoir, southwestern Chad	07.08.2002
	86	514GG/P	SB1025	4 3 4 4* 3 3.1	deleted	absent	EU887557	N'Djaména Abattoir, southwestern Chad	07.08.2002
micro-array	87	515PM/P	SB1102	X X 5 4* 4 3.1	intact	present	seq Af1 intact	N'Djaména Abattoir, southwestern Chad	07.08.2002
	88	516GG/P	SB1025	4 3 4 4* 3 3.1	deleted	absent		N'Djaména Abattoir, southwestern Chad	07.08.2002
	89	523PM/P	SB0944	4 5 6 5* 3 3.1	deleted	absent		N'Djaména Abattoir, southwestern Chad	08.08.2002
	90	552PM/P	SB0944	4 7* 5 4* 3 3.1	deleted	absent		N'Djaména Abattoir, southwestern Chad	21.08.2002
	91	577GG/P	SB1025	4 3 4 4* 3 3.1	deleted	absent		N'Djaména Abattoir, southwestern Chad	23.08.2002
	92	581GG/P	SB0944	4 5 6 5* 3 3.1	deleted	absent		N'Djaména Abattoir, southwestern Chad	23.08.2002
	93	583PM/P	SB1025	4 3 4 4* 3 3.1	deleted	absent		N'Djaména Abattoir, southwestern Chad	23.08.2002
	94	584GG/P	SB1025	4 3 4 4* 3 3.1	deleted	absent		N'Djaména Abattoir, southwestern Chad	23.08.2002
	95	588PM/P	SB1025	4 X X X 3 X	deleted	absent		N'Djaména Abattoir, southwestern Chad	24.08.2002
	96	550GG/P	SB0944	4 5 5 2 3 3.1	deleted	absent		N'Djaména Abattoir, southwestern Chad	21.08.2002
	953	542	SB0944	ND	deleted	absent		N'Djaména Abattoir, southwestern Chad	16.07.2002
	957	633	SB0944	ND	deleted	absent		N'Djaména Abattoir, southwestern Chad	28.08.2002
	977	635	SB0944	4 8 5 3* 3 3.1	deleted	absent		N'Djaména Abattoir, southwestern Chad	28.08.2002
	978	648	SB0944	ND	deleted	absent		N'Djaména Abattoir, southwestern Chad	27.08.2002

Chad, Sarh	13 HPS pm P	SB0944	X 2 X 4* 3 3.1	ND	absent		Sarh abattoir, southern Chad	July-November 2005
n = 24	18 HPS pm G	SB0951	4 6 5 4* 3 3.1	deleted	absent	EU887548	Sarh abattoir, southern Chad	July-November 2005
	19 HPS gg prescap P	SB0944	3 X 5 4* 3 3.1	deleted	absent		Sarh abattoir, southern Chad	July-November 2005
	208 gg prescap G	SB1099	4 5 5 X 3 3.1	deleted	absent	EU887547	Sarh abattoir, southern Chad	July-November 2005
	208 pm P	SB0944	3 5 5 4* 3 3.1	deleted	absent		Sarh abattoir, southern Chad	July-November 2005
	28 HPS gg mam P	SB1025	ND	deleted	absent		Sarh abattoir, southern Chad	July-November 2005
	30 HPS pm P	SB0944	4 4 5 X 3 3.1	deleted	absent		Sarh abattoir, southern Chad	July-November 2005
	322 gg prescap P	SB1453	X X X 4* 3 3.1	deleted	absent		Sarh abattoir, southern Chad	July-November 2005
	368 pm P	SB1104	X X 5 4* 3 3.1	deleted	absent		Sarh abattoir, southern Chad	July-November 2005
	43 HPS gg prescap P	SB1025	ND	deleted	absent		Sarh abattoir, southern Chad	July-November 2005
	526 gg prescap G	SB1454	4 X 6 4* 3 3.1	deleted	absent	EU887546	Sarh abattoir, southern Chad	July-November 2005
	526 gg prescap P	SB0850	4 X 6 4* 3 3.1	deleted	absent	EU887545	Sarh abattoir, southern Chad	July-November 2005
	54 HPS gg prescap G	SB0944	4 5 5 4* 3 3.1	deleted	absent		Sarh abattoir, southern Chad	July-November 2005
	57 HPS pm P	SB1025	4 3 4 4* 3 3.1	deleted	absent	EU887544	Sarh abattoir, southern Chad	July-November 2005
	65 HPS gg mam P	SB1025	5 5 4 4* 3 3.1	deleted	absent		Sarh abattoir, southern Chad	July-November 2005
	656 gg mam P	SB0944	X X X X 3 3.1	deleted	absent		Sarh abattoir, southern Chad	July-November 2005
	685 gg mam G	SB1103	7 5 5 4* 3 3.1	intact	absent		Sarh abattoir, southern Chad	July-November 2005
	709 gg prescap G	SB0944	ND	ND	absent		Sarh abattoir, southern Chad	July-November 2005
	806 rein P	SB1104	2 X 5 4* 3 3.1	deleted	absent	EU887543	Sarh abattoir, southern Chad	July-November 2005
	827 gg mam P	SB1099	3 5 5 4* 3 3.1	deleted	absent		Sarh abattoir, southern Chad	July-November 2005
	853 gg mam P	SB0951	X X X 4* 3 3.1	ND	absent		Sarh abattoir, southern Chad	July-November 2005
	866 pm P	SB0951	ND	ND	absent		Sarh abattoir, southern Chad	July-November 2005
	940 gg prescap P	SB0944	ND	ND	absent		Sarh abattoir, southern Chad	July-November 2005
	950 gg mam P	SB0944	4 X 5 4* 3 3.1	deleted	absent	EU887542	Sarh abattoir, southern Chad	July-November 2005
Nigeria	4	SB0944	5 5 5 4* 3 3.1	Deleted	absent		Bodija Abattoir (Ibadan, Southwestern-Nigeria	2003
Published strains	7	SB0944	5 5 5 4* 3 3.1	Deleted	absent		Bodija Abattoir (Ibadan, Southwestern-Nigeria	2003
n = 15	10	SB0944	5 5 5 4* 3 3.1	Deleted	absent		Bodija Abattoir (Ibadan, Southwestern-Nigeria	2003
	11	SB0944	5 5 5 4* 3 3.1	Deleted	absent		Bodija Abattoir (Ibadan, Southwestern-Nigeria	2003
	21	SB0952	5 5 5 4* 3 3.1	Deleted	absent		Bodija Abattoir (Ibadan, Southwestern-Nigeria	2003
	33	SB1025	5 5 4 4* 3 3.1	Deleted	absent		Bodija Abattoir (Ibadan, Southwestern-Nigeria	2003
	34	SB0952	5 5 5 4* 3 3.1	Deleted	absent		Bodija Abattoir (Ibadan, Southwestern-Nigeria	2003
	36	SB0944	4 5 5 4* 3 3.1	Deleted	absent		Bodija Abattoir (Ibadan, Southwestern-Nigeria	2003
	37	SB0952	5 5 5 4* 3 3.1	Deleted	absent		Bodija Abattoir (Ibadan, Southwestern-Nigeria	2003
	43	SB0944	5 5 3 4* 3 3.1	Deleted	absent		Bodija Abattoir (Ibadan, Southwestern-Nigeria	2003
	44	SB1026	5 5 5 4* 3 3.1	Deleted	absent		Bodija Abattoir (Ibadan, Southwestern-Nigeria	2003
	45	SB1025	5 5 4 4* 3 3.1	Deleted	absent		Bodija Abattoir (Ibadan, Southwestern-Nigeria	2003
	47	SB0944	5 5 3 4* 3 3.1	Deleted	absent		Bodija Abattoir (Ibadan, Southwestern-Nigeria	2003
	49	SB0944	5 5 3 4* 3 3.1	Deleted	absent		Bodija Abattoir (Ibadan, Southwestern-Nigeria	2003
	54	SB0944	5 5 3 4* 3 3.1	Deleted	absent		Bodija Abattoir (Ibadan, Southwestern-Nigeria	2003

Nigeria	5	B5-05	SB1420	5 5 6 4* 3 3.1	Deleted	absent	Bodija Abattoir (Ibadan, Southwestern-Nigeria	April to August 2004
new strains	6	B6-05	SB1420	5 5 6 4* 3 3.1	Deleted	absent	Bodija Abattoir (Ibadan, Southwestern-Nigeria	April to August 2004
n = 163	175	B175-05	SB1420	5 5 6 4* 3 2.1	Deleted	absent	Bodija Abattoir (Ibadan, Southwestern-Nigeria	April to August 2004
	2	B2-05	SB1421	5 5 5 4* 3 3.1	Deleted	absent	Bodija Abattoir (Ibadan, Southwestern-Nigeria	April to August 2004
	3	B3-05	SB1421	5 5 5 4* 3 3.1	Deleted	absent	Bodija Abattoir (Ibadan, Southwestern-Nigeria	April to August 2004
	4	B4-05	SB1421	X 5 5 4* 3 3.1	ND	absent	Bodija Abattoir (Ibadan, Southwestern-Nigeria	April to August 2004
	7	B7-05	SB1421	5 5 5 4* 3 3.1	Deleted	absent	Bodija Abattoir (Ibadan, Southwestern-Nigeria	April to August 2004
	8	B8-05	SB1421	5 5 5 4* 3 3.1	Deleted	absent	Bodija Abattoir (Ibadan, Southwestern-Nigeria	April to August 2004
	80	B80-05	SB1421	5 6 5 4* 3 3.1	Deleted	absent	Bodija Abattoir (Ibadan, Southwestern-Nigeria	April to August 2004
	55	B55-05	SB1444	5 6 5 4* 3 3.1	Deleted	absent	Bodija Abattoir (Ibadan, Southwestern-Nigeria	April to August 2004
	66	B66-05	SB1444	5 5 5 4* 3 3.1	Deleted	absent	Bodija Abattoir (Ibadan, Southwestern-Nigeria	April to August 2004
	169	B169-05	SB1444	5 2 5 4* 3 3.1	Deleted	absent	Bodija Abattoir (Ibadan, Southwestern-Nigeria	April to August 2004
	170	B170-05	SB1444	5 2 5 4* 3 3.1	Deleted	absent	Bodija Abattoir (Ibadan, Southwestern-Nigeria	April to August 2004
	160	B160-05	SB1445	5 5 5 4* 3 3.1	Deleted	absent	Bodija Abattoir (Ibadan, Southwestern-Nigeria	April to August 2004
	107	B107-05	SB1445	X 5 X 4* 3 3.1	ND	absent	Bodija Abattoir (Ibadan, Southwestern-Nigeria	April to August 2004
	13	B13-05	SB1027	4 4 5 4* 3 3.1	Deleted	absent	Bodija Abattoir (Ibadan, Southwestern-Nigeria	April to August 2004
	18	B18-05	SB1027	5 5 5 4* 3 3.1	Deleted	absent	Bodija Abattoir (Ibadan, Southwestern-Nigeria	April to August 2004
	22	B22-05	SB1027	5 5 5 4* X 3.1	ND	absent	Bodija Abattoir (Ibadan, Southwestern-Nigeria	April to August 2004
	25	B25-05	SB1027	5 X 5 4* 3 3.1	ND	absent	Bodija Abattoir (Ibadan, Southwestern-Nigeria	April to August 2004
	44	B44-05	SB1027	5 5 5 4* 3 3.1	Deleted	absent	Bodija Abattoir (Ibadan, Southwestern-Nigeria	April to August 2004
	52	B52-05	SB1027	5 6 5 4* 3 3.1	Deleted	absent	Bodija Abattoir (Ibadan, Southwestern-Nigeria	April to August 2004
	54	B54-05	SB1027	5 5 5 4* 2 3.1	Deleted	absent	Bodija Abattoir (Ibadan, Southwestern-Nigeria	April to August 2004
	56	B56-05	SB1027	5 6 5 4*3 3.1	Deleted	absent	Bodija Abattoir (Ibadan, Southwestern-Nigeria	April to August 2004
	57	B57-05	SB1027	4 5 5 4* 3 3.1	Deleted	absent	Bodija Abattoir (Ibadan, Southwestern-Nigeria	April to August 2004
	65	B65-05	SB1027	5 6 5 4* 3 3.1	Deleted	absent	Bodija Abattoir (Ibadan, Southwestern-Nigeria	April to August 2004
	72	B72-05	SB1027	4 5 5 4* 3 3.1	Deleted	absent	Bodija Abattoir (Ibadan, Southwestern-Nigeria	April to August 2004
	74	B74-05	SB1027	5 5 5 4* 3 3.1	Deleted	absent	Bodija Abattoir (Ibadan, Southwestern-Nigeria	April to August 2004
	81	B81-05	SB1027	5 5 6 4* 3 3.1	Deleted	absent	Bodija Abattoir (Ibadan, Southwestern-Nigeria	April to August 2004
	82	B82-05	SB1027	5 5 6 4* 3 3.1	Deleted	absent	Bodija Abattoir (Ibadan, Southwestern-Nigeria	April to August 2004
	88	B88-05	SB1027	5 5 5 4* 3 3.1	Deleted	absent	Bodija Abattoir (Ibadan, Southwestern-Nigeria	April to August 2004
	94	B94-05	SB1027	4 2 5 4* 3 3.1	Deleted	absent	Bodija Abattoir (Ibadan, Southwestern-Nigeria	April to August 2004
	101	B101-05	SB1027	5 5 5 4* 3 3.1	Deleted	absent	Bodija Abattoir (Ibadan, Southwestern-Nigeria	April to August 2004
	105	B105-05	SB1027	5 5 5 4* 3 3.1	Deleted	absent	Bodija Abattoir (Ibadan, Southwestern-Nigeria	April to August 2004
	108	B108-05	SB1027	5 5 6 4* 3 3.1	Deleted	absent	Bodija Abattoir (Ibadan, Southwestern-Nigeria	April to August 2004
	111	B111-05	SB1027	5 6 5 4* 3 3.1	Deleted	absent	Bodija Abattoir (Ibadan, Southwestern-Nigeria	April to August 2004
	122	B122-05	SB1027	ND	ND	absent	Bodija Abattoir (Ibadan, Southwestern-Nigeria	April to August 2004
	123	B123-05	SB1027	5 6 5 4* 3 3.1	Deleted	absent	Bodija Abattoir (Ibadan, Southwestern-Nigeria	April to August 2004
	155	B155-05	SB1027	5 5 5 4* 3 3.1	Deleted	absent	Bodija Abattoir (Ibadan, Southwestern-Nigeria	April to August 2004
	159	B159-05	SB1027	5 5 6 4* 3 3.1	Deleted	absent	Bodija Abattoir (Ibadan, Southwestern-Nigeria	April to August 2004
	165	B165-05	SB1027	5 5 6 4* 3 3.1	Deleted	absent	Bodija Abattoir (Ibadan, Southwestern-Nigeria	April to August 2004
	26	B26-05	SB1025	5 5 4 4* 3 3.1	Deleted	absent	Bodija Abattoir (Ibadan, Southwestern-Nigeria	April to August 2004
	29	B29-05	SB1025	5 5 4 4* 3 3.1	Deleted	absent	Bodija Abattoir (Ibadan, Southwestern-Nigeria	April to August 2004
	42	B42-05	SB1025	5 5 4 4* 3 3.1	Deleted	absent	Bodija Abattoir (Ibadan, Southwestern-Nigeria	April to August 2004
	43	B43-05	SB1025	5 5 4 4* 3 3.1	Deleted	absent	Bodija Abattoir (Ibadan, Southwestern-Nigeria	April to August 2004
	117	B117-05	SB1025	4 3 4 4* 3 3.1	Deleted	absent	Bodija Abattoir (Ibadan, Southwestern-Nigeria	April to August 2004
	144	B144-05	SB1025	5 5 4 4* 3 3.1	Deleted	absent	Bodija Abattoir (Ibadan, Southwestern-Nigeria	April to August 2004
	103	B103-05	SB1443	X 5 5 4* 3 3.1	ND	absent	Bodija Abattoir (Ibadan, Southwestern-Nigeria	April to August 2004

180	B180-05	SB1422	X 5 3 4* 3 3.1	Deleted	absent		Bodija Abattoir (Ibadan, Southwestern-Nigeria	April to August 2004
172	B172-05	SB1423	5 5 5 4* 3 3.1	Deleted	absent		Bodija Abattoir (Ibadan, Southwestern-Nigeria	April to August 2004
167	B167-05	SB1424	5 5 5 4* 3 3.1	Deleted	absent		Bodija Abattoir (Ibadan, Southwestern-Nigeria	April to August 2004
179	B179-05	SB1424	5 5 5 4* 3 3.1	Deleted	absent		Bodija Abattoir (Ibadan, Southwestern-Nigeria	April to August 2004
162	B162-05	SB1425	5 5 3 4* 3 3.1	Deleted	absent		Bodija Abattoir (Ibadan, Southwestern-Nigeria	April to August 2004
127	B127-05	SB1441	5 5 X 4* 3 3.1	ND	absent		Bodija Abattoir (Ibadan, Southwestern-Nigeria	April to August 2004
134	B134-05	SB1442	4 5 5 4* 3 3.1	Deleted	absent		Bodija Abattoir (Ibadan, Southwestern-Nigeria	April to August 2004
112	B112-05	SB1426	ND	Deleted	absent		Bodija Abattoir (Ibadan, Southwestern-Nigeria	April to August 2004
154	B154-05	SB1427	5 5 5 4* 3 3.1	Deleted	absent		Bodija Abattoir (Ibadan, Southwestern-Nigeria	April to August 2004
136	B136-05	SB1428	ND	Deleted	absent		Bodija Abattoir (Ibadan, Southwestern-Nigeria	April to August 2004
121	B121-05	SB1429	5 5 5 4* 3 3.1	Deleted	absent		Bodija Abattoir (Ibadan, Southwestern-Nigeria	April to August 2004
119	B119-05	SB1430	X 5 3 4* 3 3.1	Deleted	absent		Bodija Abattoir (Ibadan, Southwestern-Nigeria	April to August 2004
120	B120-05	SB1430	5 5 3 4* 3 3.1	Deleted	absent		Bodija Abattoir (Ibadan, Southwestern-Nigeria	April to August 2004
100	B100-05	SB1431	5 5 4 4* 3 3.1	Deleted	absent		Bodija Abattoir (Ibadan, Southwestern-Nigeria	April to August 2004
1	B1-05	SB1099	4 5 5 4* 3 3.1	Deleted	absent	poor seq Af1 deleted	Bodija Abattoir (Ibadan, Southwestern-Nigeria	April to August 2004
99	B99-05	SB1099	4 X 5 4* 3 X	ND	absent		Bodija Abattoir (Ibadan, Southwestern-Nigeria	April to August 2004
20	B20-05	SB1432	5 5 5 4* 3 3.1	Deleted	absent		Bodija Abattoir (Ibadan, Southwestern-Nigeria	April to August 2004
87	B87-05	SB1432	5 5 5 4* 3 3.1	Deleted	absent		Bodija Abattoir (Ibadan, Southwestern-Nigeria	April to August 2004
93	B93-05	SB1432	5 5 5 4* 3 3.1	Deleted	absent		Bodija Abattoir (Ibadan, Southwestern-Nigeria	April to August 2004
115	B115-05	SB1432	5 5 5 4* 3 3.1	Deleted	absent		Bodija Abattoir (Ibadan, Southwestern-Nigeria	April to August 2004
79	B79-05	SB1433	5 5 5 4* 3 3.1	Deleted	absent		Bodija Abattoir (Ibadan, Southwestern-Nigeria	April to August 2004
39	B39-05	SB0951	5 3 5 4* 3 3.1	Deleted	absent		Bodija Abattoir (Ibadan, Southwestern-Nigeria	April to August 2004
163	B163-05	SB0951	5 3 5 4* 3 3.1	Deleted	absent		Bodija Abattoir (Ibadan, Southwestern-Nigeria	April to August 2004
171	B171-05	SB0951	5 5 5 4* 3 3.1	Deleted	absent		Bodija Abattoir (Ibadan, Southwestern-Nigeria	April to August 2004
32	B32-05	SB1434	5 5 5 4* 3 3.1	Deleted	absent		Bodija Abattoir (Ibadan, Southwestern-Nigeria	April to August 2004
24	B24-05	SB1437	5 5 5 4* 3 3.1	Deleted	absent		Bodija Abattoir (Ibadan, Southwestern-Nigeria	April to August 2004
21	B21-05	SB1436	5 5 5 4* 4 3.1	Deleted	absent		Bodija Abattoir (Ibadan, Southwestern-Nigeria	April to August 2004
19	B19-05	SB1435	5 5 4 4* 3 3.1	Deleted	absent		Bodija Abattoir (Ibadan, Southwestern-Nigeria	April to August 2004
89	B89-05	SB1435	5 5 4 4* 3 3.1	Deleted	absent		Bodija Abattoir (Ibadan, Southwestern-Nigeria	April to August 2004
11	B11-05	SB0328	4 4 3 4* 3 3.1	Deleted	absent		Bodija Abattoir (Ibadan, Southwestern-Nigeria	April to August 2004
128	B128-05	SB0328	4 4 3 4* 3 3.1	Deleted	absent		Bodija Abattoir (Ibadan, Southwestern-Nigeria	April to August 2004
130	B130-05	SB0328	ND	ND	absent		Bodija Abattoir (Ibadan, Southwestern-Nigeria	April to August 2004
131	B131-05	SB0328	4 5 3 4* 2 4.1	Deleted	absent		Bodija Abattoir (Ibadan, Southwestern-Nigeria	April to August 2004
176	B176-05	SB0328	3 4 3 4* 3 3.1	Deleted	absent		Bodija Abattoir (Ibadan, Southwestern-Nigeria	April to August 2004
9	B9-05	SB1438	5 5 3 4* 3 3.1	Deleted	absent		Bodija Abattoir (Ibadan, Southwestern-Nigeria	April to August 2004
137	B137-05	SB1438	5 5 3 4* 3 3.1	Deleted	absent		Bodija Abattoir (Ibadan, Southwestern-Nigeria	April to August 2004
133	B133-05	SB1439	5 5 3 3* 3 3.1	Deleted	absent		Bodija Abattoir (Ibadan, Southwestern-Nigeria	April to August 2004
141	B141-05	SB1439	5 5 3 4* 3 3.1	Deleted	absent		Bodija Abattoir (Ibadan, Southwestern-Nigeria	April to August 2004
147	B147-05	SB1439	5 5 5 4* 3 3.1	Deleted	absent		Bodija Abattoir (Ibadan, Southwestern-Nigeria	April to August 2004
70	B70-05	SB1440	5 5 5 4* 3 3.1	Deleted	absent		Bodija Abattoir (Ibadan, Southwestern-Nigeria	April to August 2004
129	B129-05	SB1440	5 6 5 4* 3 3.1	Deleted	absent		Bodija Abattoir (Ibadan, Southwestern-Nigeria	April to August 2004
149	B149-05	SB1440	5 5 5 4* 3 3.1	Deleted	absent		Bodija Abattoir (Ibadan, Southwestern-Nigeria	April to August 2004
156	B156-05	SB1428	5 5 5 4* 3 3.1	Deleted	absent		Bodija Abattoir (Ibadan, Southwestern-Nigeria	April to August 2004
10	B10-05	SB0944	5 5 3 X 3 3.1	ND	absent		Bodija Abattoir (Ibadan, Southwestern-Nigeria	April to August 2004
14	B14-05	SB0944	5 5 3 4* 3 3.1	Deleted	absent		Bodija Abattoir (Ibadan, Southwestern-Nigeria	April to August 2004
15	B15-05	SB0944	5 5 3 4* 3 3.1	Deleted	absent	EU887555	Bodija Abattoir (Ibadan, Southwestern-Nigeria	April to August 2004

16	B16-05	SB0944	5 5 3 4* 3 3.1	Deleted	absent	Bodija Abattoir (Ibadan, Southwestern-Nigeria	April to August 2004
17	B17-05	SB0944	5 5 3 4* 3 3.1	Deleted	absent	Bodija Abattoir (Ibadan, Southwestern-Nigeria	April to August 2004
27	B27-05	SB0944	5 4 5 4* 3 3.1	Deleted	absent	Bodija Abattoir (Ibadan, Southwestern-Nigeria	April to August 2004
28	B28-05	SB0944	4 5 5 4* 3 3.1	Deleted	absent	Bodija Abattoir (Ibadan, Southwestern-Nigeria	April to August 2004
30	B30-05	SB0944	5 5 3 4* 3 3.1	Deleted	absent	Bodija Abattoir (Ibadan, Southwestern-Nigeria	April to August 2004
31	B31-05	SB0944	5 5 3 4* 3 3.1	Deleted	absent	Bodija Abattoir (Ibadan, Southwestern-Nigeria	April to August 2004
34	B34-05	SB0944	5 5 5 4* 3 3.1	Deleted	absent	Bodija Abattoir (Ibadan, Southwestern-Nigeria	April to August 2004
35	B35-05	SB0944	5 5 3 4* 3 3.1	Deleted	absent	Bodija Abattoir (Ibadan, Southwestern-Nigeria	April to August 2004
36	B36-05	SB0944	5 5 5 4* 3 3.1	Deleted	absent	Bodija Abattoir (Ibadan, Southwestern-Nigeria	April to August 2004
37	B37-05	SB0944	5 5 5 4* 3 3.1	Deleted	absent	Bodija Abattoir (Ibadan, Southwestern-Nigeria	April to August 2004
38	B38-05	SB0944	5 4 X 4* 3 3.1	ND	absent	Bodija Abattoir (Ibadan, Southwestern-Nigeria	April to August 2004
40	B40-05	SB0944	5 5 5 4* 3 3.1	Deleted	absent	Bodija Abattoir (Ibadan, Southwestern-Nigeria	April to August 2004
41	B41-05	SB0944	6 4 5 4* 3 3.1	Deleted	absent	Bodija Abattoir (Ibadan, Southwestern-Nigeria	April to August 2004
45	B45-05	SB0944	5 4 6 4* 3 3.1	Deleted	absent	Bodija Abattoir (Ibadan, Southwestern-Nigeria	April to August 2004
46	B46-05	SB0944	ND	Deleted	absent	Bodija Abattoir (Ibadan, Southwestern-Nigeria	April to August 2004
50	B50-05	SB0944	5 6 5 4* 3 3.1	Deleted	absent	Bodija Abattoir (Ibadan, Southwestern-Nigeria	April to August 2004
51	B51-05	SB0944	5 X 5 4* 3 3.1	ND	absent	Bodija Abattoir (Ibadan, Southwestern-Nigeria	April to August 2004
53	B53-05	SB0944	5 4 6 4* 3 3.1	Deleted	absent	Bodija Abattoir (Ibadan, Southwestern-Nigeria	April to August 2004
59	B59-05	SB0944	5 4 5 4* 3 1.3	Deleted	absent	Bodija Abattoir (Ibadan, Southwestern-Nigeria	April to August 2004
60	B60-05	SB0944	5 4 5 4* 3 1.3	Deleted	absent	Bodija Abattoir (Ibadan, Southwestern-Nigeria	April to August 2004
61	B61-05	SB0944	5 5 5 4* 3 3.1	Deleted	absent	Bodija Abattoir (Ibadan, Southwestern-Nigeria	April to August 2004
62	B62-05	SB0944	5 5 5 4* 3 3.1	Deleted	absent	Bodija Abattoir (Ibadan, Southwestern-Nigeria	April to August 2004
63	B63-05	SB0944	5 5 3 4* 3 3.1	Deleted	absent	Bodija Abattoir (Ibadan, Southwestern-Nigeria	April to August 2004
64	B64-05	SB0944	ND	ND	absent	Bodija Abattoir (Ibadan, Southwestern-Nigeria	April to August 2004
67	B67-05	SB0944	5 5 3 4* 3 3.1	Deleted	absent	Bodija Abattoir (Ibadan, Southwestern-Nigeria	April to August 2004
68	B68-05	SB0944	5 5 5 4* 3 3.1	Deleted	absent	Bodija Abattoir (Ibadan, Southwestern-Nigeria	April to August 2004
69	B69-05	SB0944	5 X 3 4* 3 3.1	Deleted	absent	Bodija Abattoir (Ibadan, Southwestern-Nigeria	April to August 2004
71	B71-05	SB0944	5 5 3 4* 3 3.1	Deleted	absent	Bodija Abattoir (Ibadan, Southwestern-Nigeria	April to August 2004
73	B73-05	SB0944	X 5 5 4* 3 3.1	ND	absent	Bodija Abattoir (Ibadan, Southwestern-Nigeria	April to August 2004
76	B76-05	SB0944	ND	ND	absent	Bodija Abattoir (Ibadan, Southwestern-Nigeria	April to August 2004
77	B77-05	SB0944	5 5 5 4* 3 3.1	Deleted	absent	Bodija Abattoir (Ibadan, Southwestern-Nigeria	April to August 2004
78	B78-05	SB0944	5 5 3 4* 3 3.1	Deleted	absent	Bodija Abattoir (Ibadan, Southwestern-Nigeria	April to August 2004
83	B83-05	SB0944	5 5 3 4* 3 3.1	Deleted	absent	Bodija Abattoir (Ibadan, Southwestern-Nigeria	April to August 2004
84	B84-05	SB0944	5 5 5 4* 3 3.1	Deleted	absent	Bodija Abattoir (Ibadan, Southwestern-Nigeria	April to August 2004
85	B85-05	SB0944	5 5 5 4* 3 3.1	Deleted	absent	Bodija Abattoir (Ibadan, Southwestern-Nigeria	April to August 2004
86	B86-05	SB0944	ND	ND	absent	Bodija Abattoir (Ibadan, Southwestern-Nigeria	April to August 2004
90	B90-05	SB0944	5 5 X 4* 3 3.1	ND	absent	Bodija Abattoir (Ibadan, Southwestern-Nigeria	April to August 2004
91	B91-05	SB0944	5 5 X 4* 3 3.1	ND	absent	Bodija Abattoir (Ibadan, Southwestern-Nigeria	April to August 2004
92	B92-05	SB0944	6 4 7 4* 3 3.1	Deleted	absent	Bodija Abattoir (Ibadan, Southwestern-Nigeria	April to August 2004
95	B95-05	SB0944	5 5 3 4* 3 3.1	Deleted	absent	Bodija Abattoir (Ibadan, Southwestern-Nigeria	April to August 2004
96	B96-05	SB0944	X 2 X 4* 3 3.1	ND	absent	Bodija Abattoir (Ibadan, Southwestern-Nigeria	April to August 2004
97	B97-05	SB0944	5 5 3 4* 3 3.1	Deleted	absent	Bodija Abattoir (Ibadan, Southwestern-Nigeria	April to August 2004
98	B98-05	SB0944	5 4 5 4* 2 1.3	Deleted	absent	Bodija Abattoir (Ibadan, Southwestern-Nigeria	April to August 2004
104	B104-05	SB0944	X 5 X 4* 3 3.1	ND	absent	Bodija Abattoir (Ibadan, Southwestern-Nigeria	April to August 2004
106	B106-05	SB0944	5 5 6 4* 3 3.1	Deleted	absent	Bodija Abattoir (Ibadan, Southwestern-Nigeria	April to August 2004
109	B109-05	SB0944	5 5 5 4* 3 3.1	Deleted	absent	Bodija Abattoir (Ibadan, Southwestern-Nigeria	April to August 2004

	110	B110-05	SB0944	X 5 X 4* 3 3.1	ND	absent		Bodija Abattoir (Ibadan, Southwestern-Nigeria	April to August 2004
	113	B113-05	SB0944	5 5 3 4* 3 3.1	Deleted	absent		Bodija Abattoir (Ibadan, Southwestern-Nigeria	April to August 2004
	114	B114-05	SB0944	X 5 3 4* 3 3.1	ND	absent		Bodija Abattoir (Ibadan, Southwestern-Nigeria	April to August 2004
	116	B116-05	SB0944	5 5 X 4* 3 3.1	ND	absent		Bodija Abattoir (Ibadan, Southwestern-Nigeria	April to August 2004
	125	B125-05	SB0944	5 5 3 4* 3 3.1	Deleted	absent		Bodija Abattoir (Ibadan, Southwestern-Nigeria	April to August 2004
	126	B126-05	SB0944	5 5 X 4* 3 3.1	ND	absent		Bodija Abattoir (Ibadan, Southwestern-Nigeria	April to August 2004
	132	B132-05	SB0944	X X X 4* 3 3.1	ND	absent		Bodija Abattoir (Ibadan, Southwestern-Nigeria	April to August 2004
	135	B135-05	SB0944	5 5 3 4* 3 1.3	Deleted	absent		Bodija Abattoir (Ibadan, Southwestern-Nigeria	April to August 2004
	138	B138-05	SB0944	5 5 5 4* 3 3.1	Deleted	absent		Bodija Abattoir (Ibadan, Southwestern-Nigeria	April to August 2004
	140	B140-05	SB0944	5 5 4 4* 3 3.1	Deleted	absent		Bodija Abattoir (Ibadan, Southwestern-Nigeria	April to August 2004
	142	B142-05	SB0944	4 4 3 4* 3 3.1	Deleted	absent		Bodija Abattoir (Ibadan, Southwestern-Nigeria	April to August 2004
	145	B145-05	SB0944	5 5 3 4* 3 3.1	Deleted	absent		Bodija Abattoir (Ibadan, Southwestern-Nigeria	April to August 2004
	148	B148-05	SB0944	5 5 5 4* 3 3.1	Deleted	absent		Bodija Abattoir (Ibadan, Southwestern-Nigeria	April to August 2004
	151	B151-05	SB0944	X 5 X 4* X 3.1	ND	absent		Bodija Abattoir (Ibadan, Southwestern-Nigeria	April to August 2004
	152	B152-05	SB0944	5 5 5 4* 3 3.1	Deleted	absent		Bodija Abattoir (Ibadan, Southwestern-Nigeria	April to August 2004
	153	B153-05	SB0944	5 5 3 4* 3 3.1	Deleted	absent		Bodija Abattoir (Ibadan, Southwestern-Nigeria	April to August 2004
	157	B157-05	SB0944	3 5 5 4* 3 3.1	Deleted	absent		Bodija Abattoir (Ibadan, Southwestern-Nigeria	April to August 2004
	164	B164-05	SB0944	5 3 5 4* 3 3.1	Deleted	absent		Bodija Abattoir (Ibadan, Southwestern-Nigeria	April to August 2004
	168	B168-05	SB0944	5 5 3 4* X 3.1	ND	absent		Bodija Abattoir (Ibadan, Southwestern-Nigeria	April to August 2004
	173	B173-05	SB0944	ND	ND	absent		Bodija Abattoir (Ibadan, Southwestern-Nigeria	April to August 2004
	177	B177-05	SB0944	5 5 5 4* 3 3.1	Deleted	absent		Bodija Abattoir (Ibadan, Southwestern-Nigeria	April to August 2004
	178	B178-05	SB0944	5 5 3 4* 3 1.3	Deleted	absent		Bodija Abattoir (Ibadan, Southwestern-Nigeria	April to August 2004
	161		SB0944	ND	ND	absent		Bodija Abattoir (Ibadan, Southwestern-Nigeria	April to August 2004
	58		SB0944	ND	ND	absent		Bodija Abattoir (Ibadan, Southwestern-Nigeria	April to August 2004
Cameroon		55N	SB0951	4 6 5 4* 3 3.1	Deleted	absent			
n = 17		54N	SB0951	4 6 5 4* 3 3.1	Deleted	absent			
		27N	SB0944	5 5 3 4* 3 3.1	Deleted	absent			
		130	SB0944	5 3 4 4* 4 2.1	Deleted	absent			
		128	SB0944	5 5 5 4* 3 2.1	Deleted	absent	EU887553		
		164	SB0944	5 5 3 4* 3 3.1	Deleted	absent			
		165	SB0944	5 5 5 4* 3 2.1	Deleted	absent			
		151	SB0952	5 5 5 4* 3 3.1	Deleted	absent			
		96	SB1419	5 5 5 4* 3 3.1	Deleted	absent			
		19N	SB0952	5 5 6 4* 3 3.1	Deleted	absent			
		108	SB0952	5 5 5 4* 3 3.1	Deleted	absent			
		66G	SB0944	5 5 5 4* 3 2.1	Deleted	absent			
		259	SB0951	3/4 6 5 4* 3 3.1	Deleted	absent			
		3438	SB0953	5 5 5 4* 3 3.1	Deleted	absent	EU887554		
		3442	SB0953	5 5 5 4* 3 3.1	Deleted	absent			
		3470	SB0953	5 5 5 4* 3 3.1	Deleted	absent			
		3441	SB0953	2 5 5 4* 3 3.1	Deleted	absent			

Mali		06b	SB0134	6 5 5 4* 3 3.1	Intact	present		Bamako abattoir, southwestern Mali	March, April 2007
n = 20		11b	SB0300	5 2 5 4* 3 3.1	Deleted	absent	EU887538	Bamako abattoir, southwestern Mali	March, April 2007
		17b	SB1411	7 4 5 4* 3 3.1	Deleted	absent	EU887539	Bamako abattoir, southwestern Mali	March, April 2007
		18b	SB0991	7 5 5 4* 3 3.1	Intact	present		Bamako abattoir, southwestern Mali	March, April 2007
		24b	SB1412	3 4 5 4* 3 3.1	Deleted	absent	EU887540	Bamako abattoir, southwestern Mali	March, April 2007
		32b	SB0134	7 5 5 4* 3 3.1	Intact	present		Bamako abattoir, southwestern Mali	March, April 2007
		33b	SB0300	5 5 5 4* 3 3.1	Deleted	absent		Bamako abattoir, southwestern Mali	March, April 2007
		35b	SB0300	5 5 5 4* 3 3.1	Deleted	absent		Bamako abattoir, southwestern Mali	March, April 2007
		37b	SB0300	5 2 5 4* 3 3.1	Deleted	absent		Bamako abattoir, southwestern Mali	March, April 2007
		38b	SB0300	5 5 3 4* 3 3.1	Deleted	absent		Bamako abattoir, southwestern Mali	March, April 2007
		40b	SB0134	7 5 5 4* 3 3.1	Intact	present		Bamako abattoir, southwestern Mali	March, April 2007
		47b	SB0134	5 5 5 4* 3 3.1	Intact	present		Bamako abattoir, southwestern Mali	March, April 2007
		48b	SB0300	5 5 3/5 4* 3 3.1	Deleted	absent		Bamako abattoir, southwestern Mali	March, April 2007
		51b	SB0134	6 5 5 4* 3 3.1	Intact	present		Bamako abattoir, southwestern Mali	March, April 2007
		52b	SB0944	4 5 5 4* 3 3.1	Deleted	absent	EU887552	Bamako abattoir, southwestern Mali	March, April 2007
		53b	SB1410	5 6 5 4* 3 3.1	Deleted	absent	EU887541	Bamako abattoir, southwestern Mali	March, April 2007
		54b	SB1410	5 6 5 4* 3 3.1	Deleted	absent		Bamako abattoir, southwestern Mali	March, April 2007
		57b	SB0300	5 5 5 4* 3 3.1	Deleted	absent		Bamako abattoir, southwestern Mali	March, April 2007
		58b	SB0300	5 2 5 4* 3 3.1	Deleted	absent		Bamako abattoir, southwestern Mali	March, April 2007
		60b	SB0134	7 5 5 4* 3 3.1	Intact	present		Bamako abattoir, southwestern Mali	March, April 2007
Uganda		B2	SB1464		Intact	present		Kalerwe Abattoir, North of Kampala	Sep 04
n = 13		B3	SB1464		Intact	present		Kalerwe Abattoir, North of Kampala	Sep 04
		B5	SB1405		Intact	present		Kalerwe Abattoir, North of Kampala	Nov 04
		B6	SB1468		Intact	present		Kalerwe Abattoir, North of Kampala	Dez 04
		B7	SB1471		Intact	present		Kalerwe Abattoir, North of Kampala	Dez 04
		B8	SB1470		Intact	present		Kalerwe Abattoir, North of Kampala	Dec-04
		B9	SB1470		Intact	present		Kalerwe Abattoir, North of Kampala	Dec-04
		B11	SB1470		Intact	present		Kalerwe Abattoir, North of Kampala	Dec-04
		B12	SB1470		Intact	present		Kalerwe Abattoir, North of Kampala	Dec-04
		B13	SB1470		Intact	present		Kalerwe Abattoir, North of Kampala	Feb-05
		B15	SB1467		Intact	present		Kalerwe Abattoir, North of Kampala	March-05
		B16	SB1467		Intact	present		Kalerwe Abattoir, North of Kampala	March-05
		B17	SB1469		Intact	present		Kalerwe Abattoir, North of Kampala	April -05
Burundi	940130	53	SB0303		Intact	present		Slaughterhouse Bujumbura	Sep. 1993
n = 10	940133	54	SB0304		Intact	present		Slaughterhouse Bujumbura	Sep. 1993
	940326	55	SB0303		Intact	present		Gatumba	Dec. 1993
	940139	56	SB0303		Intact	present		Gatumba	Dec. 1993
	940329	57	SB0303		Intact	present		Gatumba	Dec. 1993
	940437	58	SB1388		Intact	present		Slaughterhouse Bujumbura	Jan. 1994
	940439	59	SB1388		Intact	present		Slaughterhouse Bujumbura	Jan. 1994
	940440	60	SB1388		Intact	present		Slaughterhouse Bujumbura	Jan. 1994
	940443	62	SB1388		Intact	present		Slaughterhouse Bujumbura	Mar. 1994
	940824	68	SB0303		Intact	present		Maramvya	Mai 94

South Africa	213D	SB0121		Intact	present	various	1993 - 2003
n = 11	268	SB0265		Intact	present	various	1993 - 2003
	392/94	SB0130		Intact	present	various	1993 - 2003
	251G	SB0163		Intact	absent	various	1993 - 2003
	251E	SB0163		Intact	absent	various	1993 - 2003
	781	SB1163		Intact	present	various	1993 - 2003
	1307	ND		Intact	ND	various	1993 - 2003
	1497	SB0121		Intact	present	various	1993 - 2003
	4572	SB0130		Intact	present	various	1993 - 2003
	4524	SB0121		Intact	present	various	1993 - 2003
	1502	SB0131		Intact	present	various	1993 - 2003
Madagascar	83B1	SB0136		intact	present	various	July, 1996
n = 8	27B3	SB1157		intact	present	various	June, 1997
	34B3	SB1161		intact	present	various	June, 1997
	44B3	SB0888		intact	present	various	June, 1997
	84B3	SB1160		intact	present	various	July, 1997
	85B3	SB1153		intact	present	various	July, 1997
	129B3	SB1159		intact	present	various	July, 1997
	246B3	SB1161		intact	present	various	November, 1997
Tanzania	1	SB0133	3 2 5 4* 3	intact	present	Morogoro abattoir, eastern Tanzania	
n = 14	11	SB0133	3 2 5 4* 3	intact	present	Morogoro abattoir, eastern Tanzania	
	18	SB0133	3 2 5 4* 3	intact	present	Morogoro abattoir, eastern Tanzania	
	34	SB0425	4 6 5 4* 4	intact	absent	Morogoro abattoir, eastern Tanzania	
	B3	SB0133	3 2 5 4* 3	intact	present	Morogoro abattoir, eastern Tanzania	
	D3	SB0133	3 2 5 4* 3	intact	present	Morogoro abattoir, eastern Tanzania	
	E1	SB1446	X 2 5 4* 3	intact	present	Morogoro abattoir, eastern Tanzania	
	G2	SB0133	3 2 5 4* 3	intact	present	Morogoro abattoir, eastern Tanzania	
	H1	SB0425	4 5 5 4* 4	intact	absent	Morogoro abattoir, eastern Tanzania	
	J1	SB0425	4 5 5 4* 4	intact	absent	Morogoro abattoir, eastern Tanzania	
	P2	SB0133	3 2 5 4* 3	intact	present	Morogoro abattoir, eastern Tanzania	
	R1	SB0133	3 2 5 4* 3	intact	present	Morogoro abattoir, eastern Tanzania	
	6	SB0133	3 2 5 4* 3	intact	present	Morogoro abattoir, eastern Tanzania	
	S1	SB0425	4 5 5 4* 4	intact	absent	Morogoro abattoir, eastern Tanzania	

Algeria	4		SB0121	intact	present	Algiers and Blida abattoir, northern Algeria	August-November 2007
n = 23	11		SB1448	intact	absent	Algiers and Blida abattoir, northern Algeria	August-November 2007
	26		SB0134	intact	present	Algiers and Blida abattoir, northern Algeria	August-November 2007
	31		SB0120	intact	present	Algiers and Blida abattoir, northern Algeria	August-November 2007
	34		SB0120	intact	present	Algiers and Blida abattoir, northern Algeria	August-November 2007
	46		SB0120	intact	present	Algiers and Blida abattoir, northern Algeria	August-November 2007
	47		SB0331	intact	present	Algiers and Blida abattoir, northern Algeria	August-November 2007
	61		SB0120	intact	present	Algiers and Blida abattoir, northern Algeria	August-November 2007
	61		SB0120	intact	present	Algiers and Blida abattoir, northern Algeria	August-November 2007
	89		SB0162	intact	present	Algiers and Blida abattoir, northern Algeria	August-November 2007
	95		SB1447	intact	present	Algiers and Blida abattoir, northern Algeria	August-November 2007
	102		SB1450	intact	absent	Algiers and Blida abattoir, northern Algeria	August-November 2007
	114		SB0850	intact	absent	Algiers and Blida abattoir, northern Algeria	August-November 2007
	28		SB0828	intact	present	Algiers and Blida abattoir, northern Algeria	August-November 2007
	33		SB0850	intact	absent	Algiers and Blida abattoir, northern Algeria	August-November 2007
	38		SB0120	intact	present	Algiers and Blida abattoir, northern Algeria	August-November 2007
	45		SB0120	intact	present	Algiers and Blida abattoir, northern Algeria	August-November 2007
	49		SB0120	intact	present	Algiers and Blida abattoir, northern Algeria	August-November 2007
	56		SB0120	intact	present	Algiers and Blida abattoir, northern Algeria	August-November 2007
	62		SB0822	intact	present	Algiers and Blida abattoir, northern Algeria	August-November 2007
	64		SB0120	intact	present	Algiers and Blida abattoir, northern Algeria	August-November 2007
	75		SB1452	intact	present	Algiers and Blida abattoir, northern Algeria	August-November 2007
	77		SB1449	intact	present	Algiers and Blida abattoir, northern Algeria	August-November 2007
Ethiopia		BTB-422	SB1477	intact	present	Jinka abattoir, southern Ethiopia	2006
n = 15		BTB-691	SB0133	intact	present	Jinka abattoir, southern Ethiopia	2007
		BTB-746	SB0133	intact	present	Jinka abattoir, southern Ethiopia	2007
		BTB-890	SB0133	intact	present	Ghimbi abattoir, western Ethiopia	2007
		BTB-891	SB1476	intact	present	Ghimbi abattoir, western Ethiopia	2007
		BTB-895	SB1476	intact	present	Ghimbi abattoir, western Ethiopia	2007
		BTB-897	SB1476	intact	present	Ghimbi abattoir, western Ethiopia	2007
		BTB-918	SB1476	intact	present	Gondar abattoir, northern Ethiopia	2007
		BTB-919	SB1476	intact	present	Gondar abattoir, northern Ethiopia	2007
		BTB-923	SB1476	intact	present	Gondar abattoir, northern Ethiopia	2007
		BTB-1090	SB0133	intact	present	Jinka abattoir, southern Ethiopia	2007
		BTB-1091	SB0133	intact	present	Jinka abattoir, southern Ethiopia	2007
		BTB-1383	SB1176	intact	absent	Addis Ababa abattoir, central Ethiopia	2007
		BTB-1387	SB1176	intact	absent	Addis Ababa abattoir, central Ethiopia	2007
		BTB-1388	SB1176	intact	absent	Addis Ababa abattoir, central Ethiopia	2007

Mozambique	1	SB0961	intact	present	Busi District, central Mozambique	Feb. 2007
n = 20	2	SB0961	intact	present	Busi District, central Mozambique	Feb. 2007
	3	SB0961	intact	present	Busi District, central Mozambique	Feb. 2007
	4	SB0961	intact	present	Busi District, central Mozambique	Feb. 2007
	5	SB0961	intact	present	Busi District, central Mozambique	Feb. 2007
	7	SB0961	intact	present	Busi District, central Mozambique	Feb. 2007
	9	SB0961	intact	present	Busi District, central Mozambique	Feb. 2007
	10	SB0961	intact	present	Busi District, central Mozambique	Feb. 2007
	11	SB0961	intact	present	Busi District, central Mozambique	Feb. 2007
	12	SB0961	intact	present	Busi District, central Mozambique	Feb. 2007
	13	SB0961	intact	present	Busi District, central Mozambique	Feb. 2007
	14	SB0961	intact	present	Busi District, central Mozambique	Feb. 2007
	15	SB0961	intact	present	Busi District, central Mozambique	Feb. 2007
	16	SB0961	intact	present	Busi District, central Mozambique	Feb. 2007
	17	SB0961	intact	present	Busi District, central Mozambique	Feb. 2007
	468	SB0140	intact	present	Marracuene District, southern Mozambique	Nov. 2007
	660	SB0140	intact	present	Machava abattoir, Southern Mozambique (from Manhica Distric))	Nov. 2007
	1086	SB0140	intact	present	Xai-xai abattoir, Southern Mozambique	Mar. 2008
	348	SB0120	intact	present	Nampula abattoir, Northen Mozambique	Nov. 2007
	851	SB0961	intact	present	Manica abattoir, Central Mozambique	Mar.2008

ND = Not Determined
X = Not Determined

Appendix 4:

Mathematical description of Bayesian model in: "Bayesian receiver operating characteristic estimation of multiple tests for diagnosis of bovine tuberculosis in Chadian cattle"

Let Y_{ki}^d denote the diagnostic test values of the *i*th animal, *i* = 1,…,*m* with disease status *d* (*d* = 0 for non-diseased; *d* = 1 for diseased) for test *k* (*k* = 1 for SICCT; *k* = 2 for SENTRY 100; *k* = 3 for GENios Pro). We assume that the test scores of the three tests are multivariate normally distributed, that is $Y_i^d = (Y_{1i}^d, Y_{2i}^d, Y_{3i}^d)^T \sim MVN(\boldsymbol{\mu}_i^d, \Sigma^d)$ where $\boldsymbol{\mu}_i^d = (\mu_{i1}^d, \mu_{i2}^d, \mu_{i3}^d)^T$ and $\Sigma_{kl}^d = \text{cov}(Y_{ik}^d, Y_{il}^d) = \sigma_{kl}^d$ with *k, l* = 1,2,3. We write the likelihood as a product of the following factors:

$$L(Y_i;)\propto\prod_{d=0,1}\prod_{i=1}^{n_d} f\left(Y_{i3}^d \mid Y_{i2}^d, Y_{i1}^d; \mu_i^d, \{\sigma_{kl}^d\}\right) f\left(Y_{i2}^d \mid Y_{i1}^d; \mu_i^d, \{\sigma_{kl}^d\}\right) f\left(Y_{i1}^d; \mu_i^d, \{\sigma_{kl}^d\}\right),$$

which have normal distributions with

$Ef\left(Y_{i1}^d; \mu_i^d, \{\sigma_{kl}^d\}\right) = \mu_{i1}^d$ and $\text{var} f\left(Y_{i1}^d; \mu_i^d, \{\sigma_{kl}^d\}\right) = \sigma_{11}^d$,

$Ef\left(Y_{i2}^d \mid Y_{i1}^d; \mu_i^d, \{\sigma_{kl}^d\}\right) = \mu_{i2}^d + \rho_{12}^d \sqrt{\dfrac{\sigma_{22}^d}{\sigma_{11}^d}}\left(Y_{i1}^d - \mu_{i1}^d\right)$ and

$\text{var} f\left(Y_{i2}^d \mid Y_{i1}^d; \mu_i^d, \{\sigma_{kl}^d\}\right) = \sigma_{22}^d\left(1-\left(\rho_{12}^d\right)^2\right)$ and

$$Ef\left(Y_{i3}^d \mid Y_{i2}^d, Y_{i1}^d; \mu_i^d, \{\sigma_{kl}^d\}\right) = \mu_{i3}^d + \left(\sigma_{11}^d\sigma_{22}^d\left(1-\left(\rho_{12}^d\right)^2\right)\right)^{-1} \times$$

$$\times\left\{\left(\rho_{32}^d + \rho_{12}^d\rho_{31}^d\right)\sigma_{11}^d\sqrt{\sigma_{22}^d\sigma_{33}^d}\left(Y_{i2}^d - \mu_{i2}^d\right) + \left(\rho_{31}^d + \rho_{12}^d\rho_{32}^d\right)\sigma_{22}^d\sqrt{\sigma_{11}^d\sigma_{33}^d}\left(Y_{i1}^d - \mu_{i1}^d\right)\right\}$$ and

$$\text{var} f\left(Y_{i3}^d \mid Y_{i2}^d, Y_{i1}^d; \mu_i^d, \{\sigma_{kl}^d\}\right) = \sigma_{33}^d\left(1-\left(\left(\rho_{32}^d\right)^2 + \left(\rho_{31}^d\right)^2\right)\left(1-\left(\rho_{12}^d\right)^2\right)^{-1}\right).$$

ρ_{kl}^d is a correlation parameter between tests *k* and *l* and it is defined in terms of the variances as follows: $\rho_{kl}^d = \dfrac{\sigma_{kl}^d}{\sqrt{\sigma_{kk}^d}\sqrt{\sigma_{ll}^d}}$.

The AUCs for each diagnostic test *k* (*k* = 1, 2, 3) can be calculated as

$AUC_k = \Phi\left(-\frac{\mu_k^{d=0}-\mu_k^{d=1}}{\sqrt{\sigma_{kk}^{d=0}+\sigma_{kk}^{d=1}}}\right)$ with Φ being the cumulative distribution function of a standard normal variable. For identifiability we assume that $\mu_k^{d=0} < \mu_k^{d=1}$, which holds for our diagnostic tests.

In an initial model, in addition to both FPA methods and SICCT, we have taken into account the results of a number of binary tests (meat inspection, direct microscopy, culture and microscopy, PCR), assuming that they had imperfect sensitivity and specificity, with the exception of PCR, which has been considered to be 100% specific. Let π be the unknown true disease prevalence in the sampled population and p_j the observed prevalence estimated from the *j*th test (*j* = 1 for meat inspection; *j* = 2 direct microscopy; *j* = 3 for culture and microscopy; *j* = 4 for PCR). Moreover, let *Ti* be the latent variable that indicates the true disease status of the *i*th animals (1 for diseased and 0 for non-diseased animals) and let *Zij* be the observed disease status from the *j*th test. We assume that $Z_{ij} \sim Bernoulli(p_j)$, $T_i \sim Bernoulli(\pi)$ and define TT_i where $TT_i = \begin{Bmatrix} T_i \text{ if PCR result is negative} \\ 1 \text{ if PCR result is positive} \end{Bmatrix}$ as we consider positive values of PCR to be the gold standard and $p_j = S_j\pi + (1-C_j)(1-\pi)$ where S_j and C_j denote the sensitivity and specificity of test *j*, respectively.

To estimate the model parameters we formulate the model within the Bayesian framework of inference and use Markov chain Monte Carlo (MCMC) simulation for model fit. The following prior distributions were adopted for the parameters:

$\mu_k^d \sim N(0,100)$, $\tau_k^d = \frac{1}{\sigma_{kk}^d} \sim Gamma(0.01,0.01)$, $\rho_{kk}^d \sim Uniform(-1,1)$ and $\pi \sim Beta(\alpha,\beta)$. α and β can be re-written in terms of mean (m_π) and variance (σ_π^2) with $\alpha = m_\pi\left(\frac{m_\pi(1-m_\pi)}{\sigma_\pi^2}-1\right)$, $\beta = (1-m_\pi)\left(\frac{m_\pi(1-m_\pi)}{\sigma_\pi^2}-1\right)$

Based on our previous observations we have set m_π to 0.1 and σ_π^2 to 0.05.

In addition, we assume $S_j \sim Beta(\alpha_j^S,\beta_j^S), j=1,...,4$ and $C_j \sim Beta(\alpha_j^C,\beta_j^C), j=1,2,3$ and $C_4 = 1$. Again, α_j^S and β_j^S or α_j^C and β_j^C can be re-written in terms of means (m_j^S, m_j^C) and variances $[(\sigma_j^S)^2, (\sigma_j^C)^2]$. We assigned means and variances as indicated in table I. Initial means for the individual test sensitivities and specificities were based on previously published estimates [63,64,245,246,250-254,257,258].

The models were fitted in WinBUGS and convergence was achieved before 30000 iterations. Convergence was assessed informally by inspection of the ergodic averages of selected parameters.

Appendix 5:

WinBUGS code for : "Bayesian receiver operating characteristic estimation of multiple tests for diagnosis of bovine tuberculosis in Chadian cattle"

(model 2A; table 14)

```
model {
for (i in 1:929) {
  tuber[i] ~ dnorm(mu1[i],tau1[i])                # test 1
  sentry100[i] ~ dnorm(condmu[i],condtau[i])      # test 2  (test2 |test1)
  genios[i]~ dnorm(condmu1[i],condtau1[i])        # test 3  (test3 |test2,test1)
  T[i]~dbern(prev_true)                           # latent disease state

# mean and precision for the test values of test 1:
  mu1[i] <- lambda1[TT[i]]
  tau1[i] <- gamma1[TT[i]]

# conditional mean and precision for the test values of test 2:
  condmu[i] <-
lambda2[TT[i]]+rho12[TT[i]]*sqrt(gamma1[TT[i]]/gamma2[TT[i]])*(tuber[i]-
lambda1[TT[i]])
  condtau[i] <- (gamma2[TT[i]])/(1-pow(rho12[TT[i]],2))

# conditional mean and precision for the test values of test 3:
  condmu1[i] <-
lambda3[TT[i]]+(1./det[TT[i]])*((rho32[TT[i]]+rho12[TT[i]]*rho31[TT[i]])*(1./(gamm
a1[TT[i]]*sqrt(gamma2[TT[i]]*gamma2[TT[i]])))*(sentry100[i]-
lambda2[TT[i]])+(rho31[TT[i]]-
rho12[TT[i]]*rho32[TT[i]])*(1./(gamma2[TT[i]]*sqrt(gamma1[TT[i]]*gamma3[TT[i]]
)))*(tuber[i]-lambda1[TT[i]]))
  condtau1[i] <-
```

```
1./((1/gamma3[TT[i]])-
((pow(rho32[TT[i]],2)+pow(rho31[TT[i]],2))/(det[TT[i]]*gamma1[TT[i]]*gamma2[T
T[i]]*gamma3[TT[i]])))
}

# considering the 100% specificity of PCR for the latent disease status
for (i in 1:929) {
  TT[i]<-equals(truinf[i],0)*(2-T[i])+equals(truinf[i],1)
}

det[1]<-(1-pow(rho12[1],2))/(gamma1[1]*gamma2[1])
det[2]<-(1-pow(rho12[2],2))/(gamma1[2]*gamma2[2])

prev_true~dbeta(a,b)                    # prior for the true disease
prevalence
a<-(m*m*(1-m)/s) -m
b<-(a/m)-a

# prior for the mean of test 1-3 for the diseased and non-diseased population:
lambda1[1] ~ dnorm(0,0.01)I(lambda1[2],)
lambda2[1] ~ dnorm(0,0.01)I(lambda2[2],)
lambda3[1] ~ dnorm(0,0.01)
lambda1[2] ~ dnorm(0,0.01)
lambda2[2] ~ dnorm(0,0.01)
lambda3[2] ~ dnorm(0,0.01)

# prior for the correlation coefficients:
rho12[1] ~ dunif(-1,1)
rho12[2] ~ dunif(-1,1)
rho31[1] ~ dunif(-1,1)
rho31[2] ~ dunif(-1,1)
rho32[1] ~ dunif(-1,1)
rho32[2] ~ dunif(-1,1)
```

```
# prior for the precision of test 1-3 for the diseased and non-diseased population:
gamma1[1] ~ dgamma(0.01,0.01)
gamma2[1] ~ dgamma(0.01,0.01)
gamma3[1] ~ dgamma(0.01,0.01)
gamma1[2] ~ dgamma(0.01,0.01)
gamma2[2] ~ dgamma(0.01,0.01)
gamma3[2] ~ dgamma(0.01,0.01)

# variance of test 1-3 for the diseased and non-diseased population:
sigma1[1] <- 1/gamma1[1]
sigma1[2] <- 1/gamma1[2]
sigma2[1] <- 1/gamma2[1]
sigma2[2] <- 1/gamma2[2]
sigma3[1] <- 1/gamma3[1]
sigma3[2] <- 1/gamma3[2]

# AUC for each test :
AUC1 <- phi(-(lambda1[2]-lambda1[1])/sqrt(sigma1[2]+sigma1[1]))
AUC2 <- phi(-(lambda2[2]-lambda2[1])/sqrt(sigma2[2]+sigma2[1]))
AUC3 <- phi(-(lambda3[2]-lambda3[1])/sqrt(sigma3[2]+sigma3[1]))

# Sensitivities/specificities for each test for given cut-off values
Se1 <- 1-phi((2.5-lambda1[1])/sqrt(sigma1[1]))
Sp1 <- phi((2.5-lambda1[2])/sqrt(sigma1[2]))
Se1OIE <- 1-phi((4.5-lambda1[1])/sqrt(sigma1[1]))
Sp1OIE <- phi((4.5-lambda1[2])/sqrt(sigma1[2]))
Se2 <- 1-phi((15-lambda2[1])/sqrt(sigma2[1]))
Sp2 <- phi((15-lambda2[2])/sqrt(sigma2[2]))
Se3 <- 1-phi((38-lambda3[1])/sqrt(sigma3[1]))
Sp3 <- phi((38-lambda3[2])/sqrt(sigma3[2]))
}
```

Printed by Books on Demand GmbH, Norderstedt / Germany